GW01375464

Higher Grade Physics Revision Notes

N R Short

Leckie & Leckie

A Word of Advice

To get the best out of your Physics for the coming exam you have to take action. In fact you have to take five actions:-

Memorise — You can't solve problems without formulae. Each time you see this box, add the formula to your formula list then sing it, shout it, scream it - whatever it takes to get it into your head!

State — There are certain important facts you've got to remember. Make a list and keep going back to it.

Describe — You could be asked about some experiments. Try making a simple diagram and a brief note.

Understand — This is the key to making a real go of the subject. Read over, repeat, argue with friends, ask your teacher. Work hard at this one.

Solve — Practice at problem solving is essential. Don't just read the answers. Treat each problem as an exam question, then make sure you fully understand the answers. (In this book I use $g = 10\,m\,s^{-2}$ instead of $9.8\,m\,s^{-2}$ to keep the numbers easy, but watch out in the exam!)

Now, good luck with Physics, but remember, the harder you work the luckier you get!

Thanks

To Bob Neill for his helpful advice and his assistance with the proofs.

To Hamish Sanderson for all his work on cartoons and diagrams.

To Bruce Ryan for proofing and page make-up.

To Eliana and Frances for their typing and copying.

Published by Leckie & Leckie
St. Andrews
Tel. 01334 - 475656
Fax. 01334 - 477392

A CIP Catalogue record for this book is available from the British Library.

ISBN 1–898890–05–6

2nd edition

Printed by Inglis Allen Kirkcaldy

CONTENTS

1. MEASUREMENT AND ERRORS

TYPES OF ERROR

Understand

1. **Systematic errors** can be caused by some constant factor.
   It causes all your results to be too high (or too low).

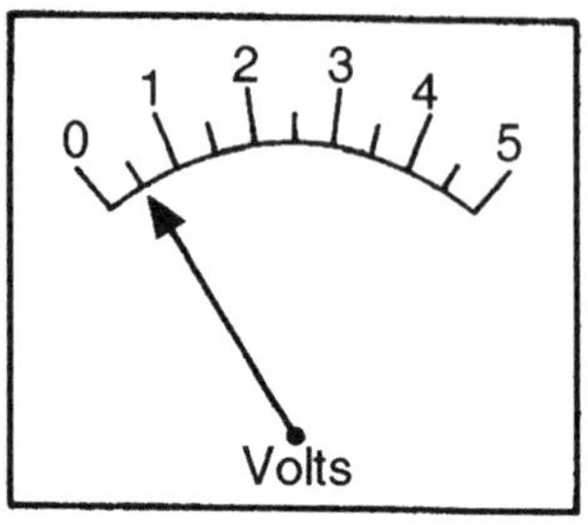

This meter is not properly set to zero.
The systematic error is 0.5 V.
All readings are 0.5 V too large.

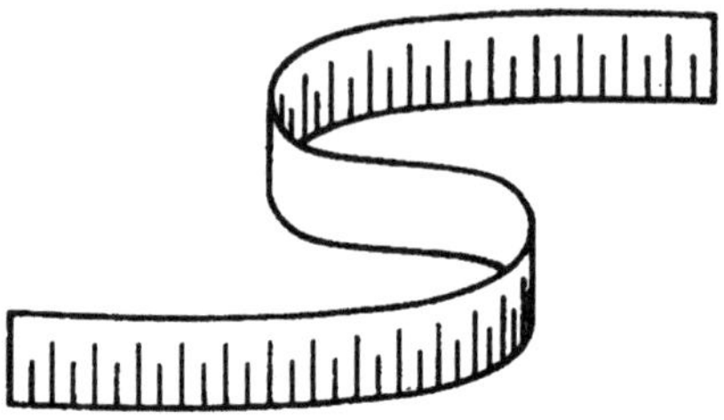

This 100 cm long measuring tape has become stretched by an extra 1 cm.
The systematic error is 1%. All readings are 1% too small. (Think about it!)

2. **Reading errors**
   On fine analogue scales an error of $\pm\frac{1}{2}$ division is usually chosen.

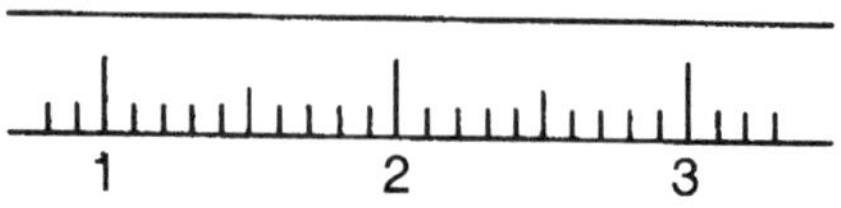

Reading error = ±0.05

On larger analogue scales $\pm\frac{1}{5}$ division may be more appropriate.

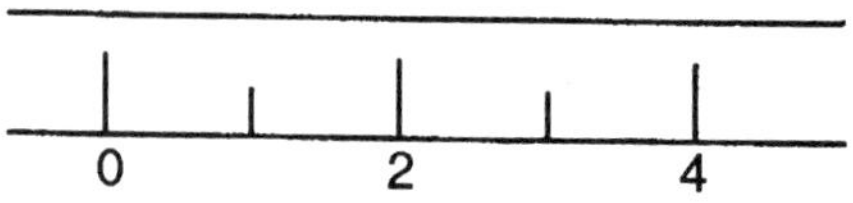

Reading error = ±0.2

On digital scales the error is ±1 on the last digit.

Reading error = ±0.01 V

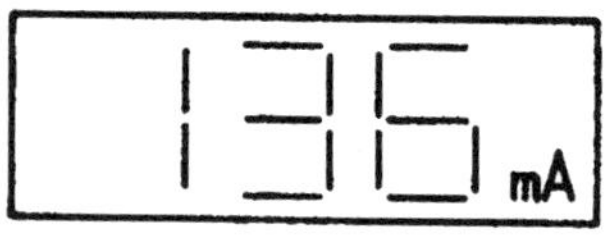

Reading error = ±1 mA

3. **Random errors** tend to show up when you make a series of measurements of the same quantity.

   Example
   Some pupils tried to measure the height of a school building.
   The results (in metres) were:- 22.0 20.1 18.0 19.1 21.3

State

4. When you have a number of measurements of the same quantity, the **mean** is a good estimate of the true value.

Example
Value of height of school building (using the figures on page 4)

$$= \frac{22.0 + 20.1 + 18.0 + 19.1 + 21.3}{5}$$

$$= \frac{100.5}{5}$$

$$= 20.1\,\text{m}$$

5. **Random error** $= \dfrac{\text{max. value} - \text{min. value}}{\text{number of readings}}$

$$= \frac{22.0 - 18.0}{5}$$

$$= 0.8\,\text{m}$$

The final result is:- Height = $20.1 \pm 0.8\,\text{m}$

6. Any error can be expressed in two ways:-

Absolute error	**Percentage error**
$100 \pm 4\,\text{cm}$	$100\,\text{cm} \pm 4\%$

7. When using a **formula** look for the quantity with the **biggest percentage error.** You can often take it as the error in your final result.

Solve

8. Use the data below to work out the percentage errors in the distance and the time. Find the speed. What is the percentage error in the speed?

Distance $s = 100 \pm 1\,\text{cm}$ Time $t = 2 \pm 0.2\,\text{s}$

ANSWERS

Distance error $= \dfrac{1}{100} \times 100$
= 1%

Time error $= \dfrac{0.2}{2} \times 100$
= 10%

Speed, $v = \dfrac{s}{t}$

$= \dfrac{100}{2}$

= 50 cm s^{-1}

% error in speed = **10%** (the bigger error).

2. MECHANICS AND THE PROPERTIES OF MATTER

VECTORS AND SCALARS

Memorise

1. Average speed = $\dfrac{\text{distance}}{\text{time}}$ $\qquad$ $v_{ave} = \dfrac{s}{t}$

2. Average velocity = $\dfrac{\text{displacement}}{\text{time}}$ $\qquad$ $\mathbf{v}_{ave} = \dfrac{\mathbf{s}}{t}$

Understand

3. Distance and speed are scalars; they have a magnitude (size) only. Displacement and velocity are vectors; they have both magnitude and direction. Look at the following examples to help you:

<u>Example 1</u>

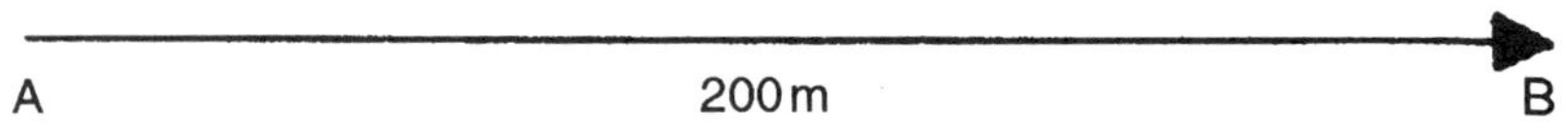

A girl runs from A to B in 40 seconds.

Distance travelled = 200 m

Displacement = 200 m East

Average speed, $v_{ave} = \dfrac{s}{t} = \dfrac{200}{40} = 5\,\text{m}\,\text{s}^{-1}$

Average velocity, $\mathbf{v}_{ave} = \dfrac{\mathbf{s}}{t} = \dfrac{200}{40} = 5\,\text{m}\,\text{s}^{-1}$ East

<u>Example 2</u>

A boy cycles round the circular track from X to Y in 20 seconds.

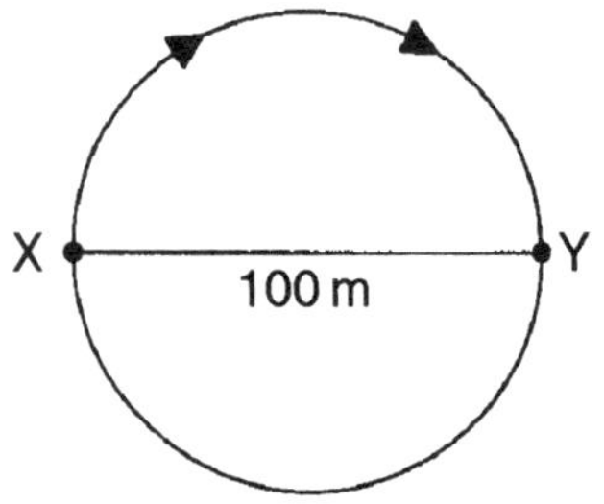

Distance travelled = 157 m (Think about it! Remember π!)

Displacement = 100 m East

Average speed, $v_{ave} = \dfrac{s}{t} = \dfrac{157}{20} = 7.85\,\text{m}\,\text{s}^{-1}$

Average velocity, $\mathbf{v}_{ave} = \dfrac{\mathbf{s}}{t} = \dfrac{100}{20} = 5\,\text{m}\,\text{s}^{-1}$ East

Understand

4. Two vectors can be combined to get a **resultant vector,** ⟶⟫.

Example 1
A girl walks 30 m North, then 40 m East.

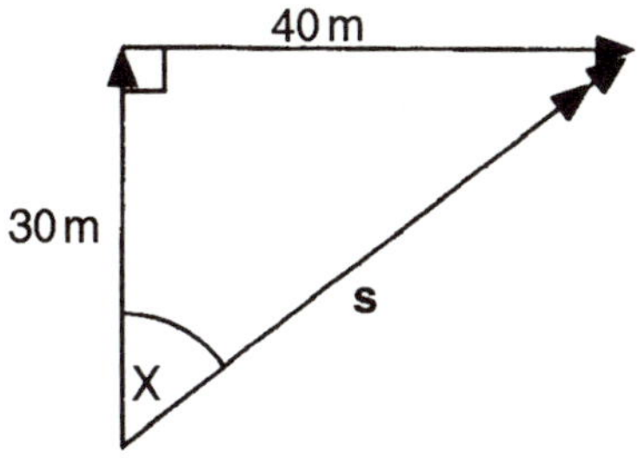

Her resultant displacement, **s**, is 50 m at an angle X of 53° East of North.

Example 2
A plane flies North at 120 m s^{-1} in a wind blowing East at 50 m s^{-1}.

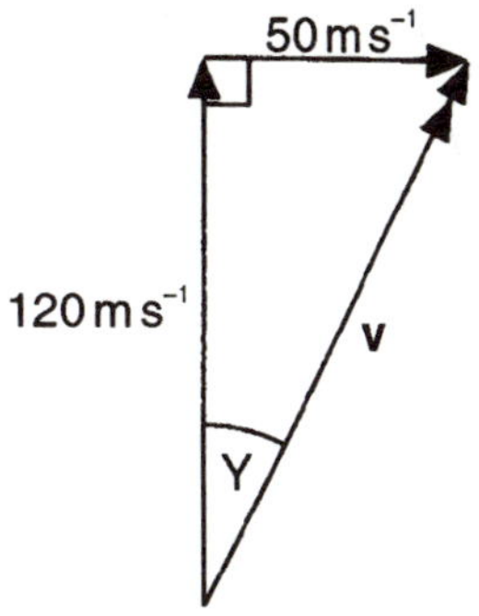

The plane's resultant velocity, **v**, is 130 ms^{-1} at an angle Y of 22.6° East of North.

5. A vector, **v**, can be resolved into two components at right angles to each other. The components are **v** sin θ and **v** cos θ.

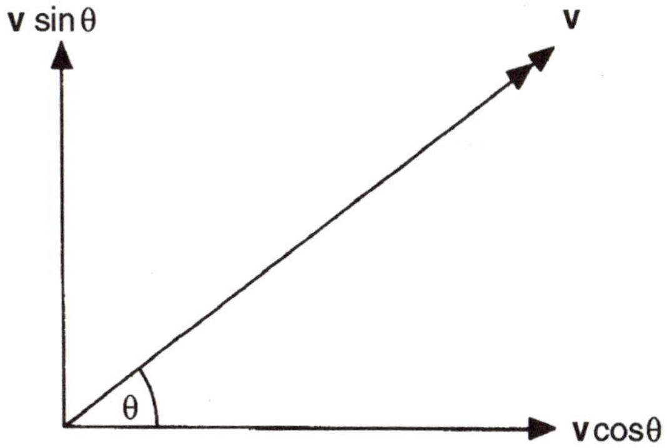

Think of this as the opposite of vector addition. It is useful when dealing with projectile motion (on page 14).

Solve

6. "Jaws" swims at 4 m s^{-1} across his pool from A to C as shown in 10 seconds. Find the pool width AB and the length AD.

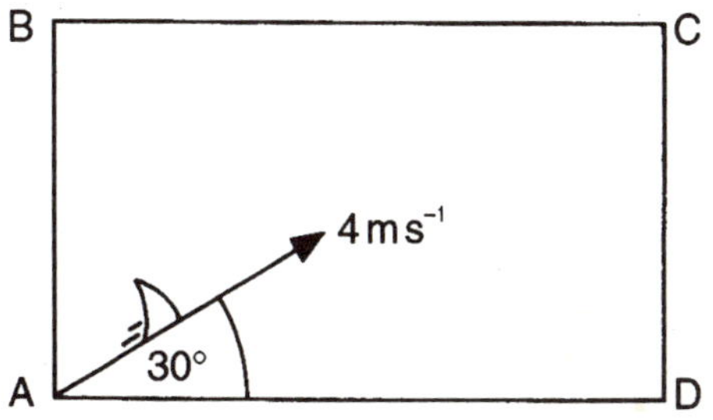

ANSWERS

In general, s = vt

⇒ AB = (v sin θ)t
= 4 × sin 30° × 10
= **20 m.**

AD = (v cos θ)t
= 4 × cos 30° × 10
= **34.6 m.**

ACCELERATION

Memorise

1. Acceleration = $\dfrac{\text{change of velocity}}{\text{time}}$

$$a = \frac{v - u}{t}$$

Units of acceleration: $m\,s^{-2}$

2. Acceleration is a vector.

Describe

3. An experiment to measure acceleration, e.g.:-

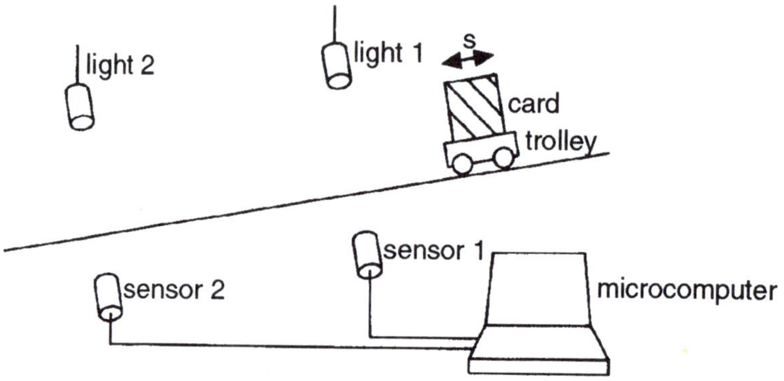

The computer measures 3 times:-

t_1 = time for card to pass through sensor 1.

t_2 = time for card to pass through sensor 2.

t_3 = time taken to go from sensor 1 to sensor 2.

Card length = s

$$u = \frac{s}{t_1} \qquad v = \frac{s}{t_2} \qquad a = \frac{v - u}{t_3}$$

Solve

4. Find the acceleration using the following data:-

Card length $s = 10\,cm$
$t_1 = 1.0\,s$
$t_2 = 0.2\,s$
$t_3 = 0.5\,s.$

You should get an answer in $cm\,s^{-2}$ equal to ten times this page number!

GRAPHS OF MOTION

Understand

1. The **shapes** of acceleration/time, velocity/time and displacement/time graphs.

These three graphs show **uniform acceleration**.

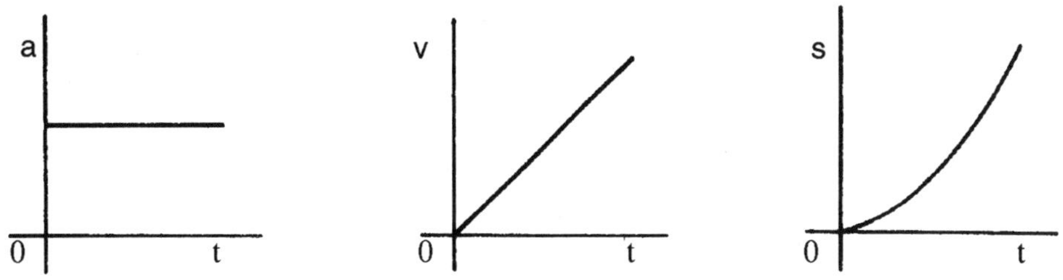

These three graphs show **constant velocity**.

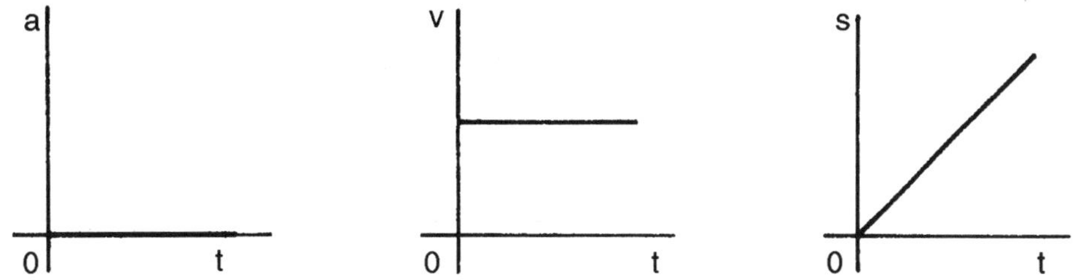

2. Graphs you might meet in the exam:

(a) A car accelerates from rest, maintains a constant velocity, then decelerates to rest.

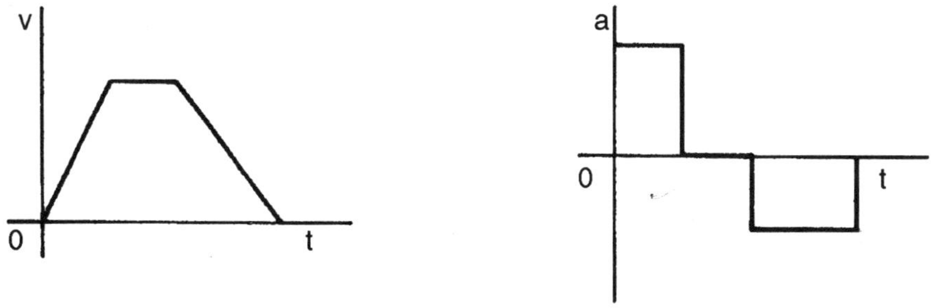

(b) A ball is thrown upwards, leaving the hand at 20 m s^{-1} and eventually returning to it.

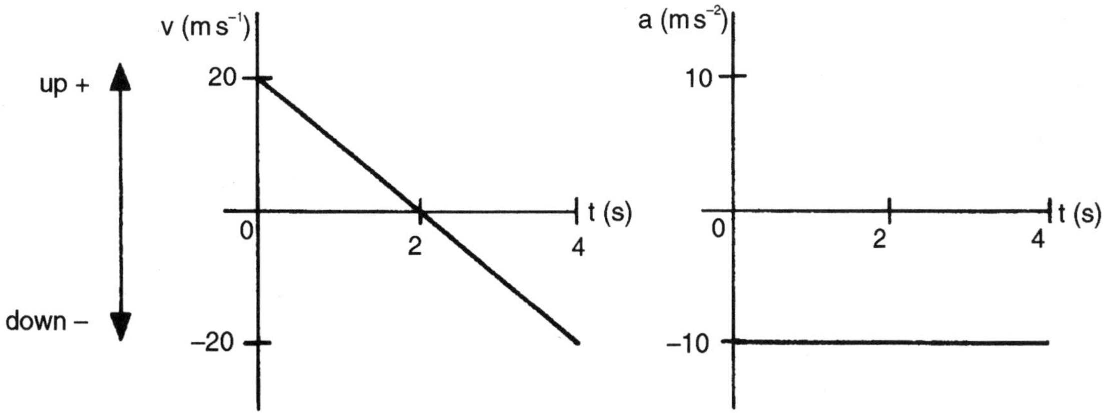

(c) A ball drops to the floor and bounces back up again.

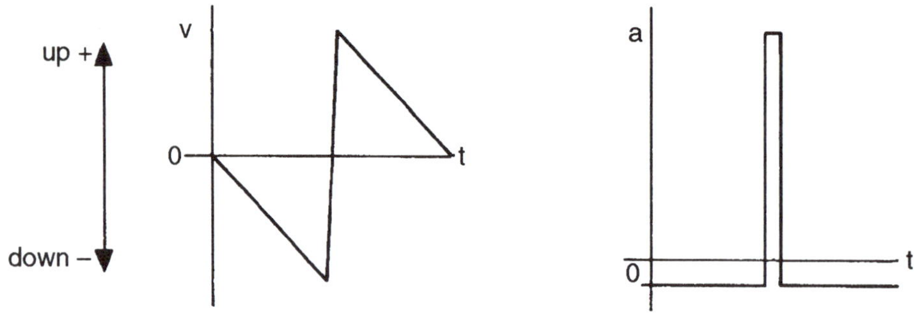

Solve

3. From the v/t graph below and the bits of help provided, try to draw the a/t graph. This type of question could well come up in the exam. Get it checked by your teacher; make the effort!

Remember, $a = \dfrac{v - u}{t}$

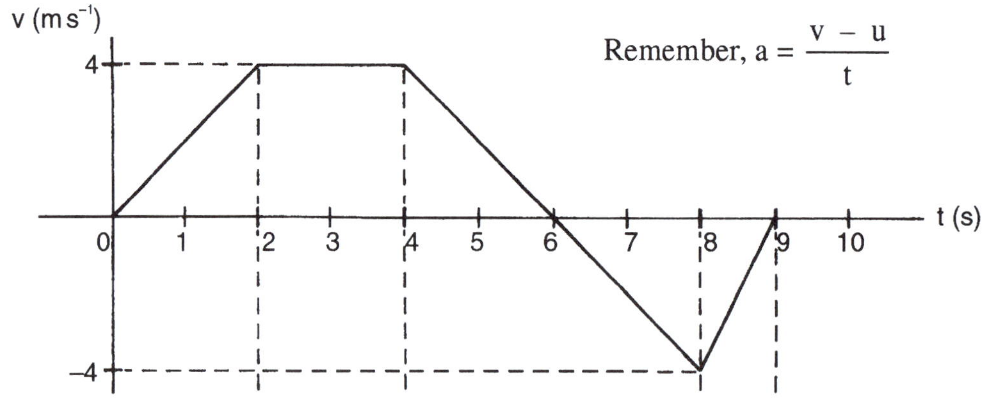

$a = \dfrac{(4) - (0)}{2}$ $a = \dfrac{(4) - (4)}{2}$ $a = \dfrac{(-4) - (4)}{4}$ $a = \dfrac{(0) - (-4)}{1}$

State

4. The **area** under a velocity/time graph gives the **displacement**.

Solve

5. Show that the displacement is the same for graphs A and B but that the distance travelled in B is five times bigger than the distance in A.

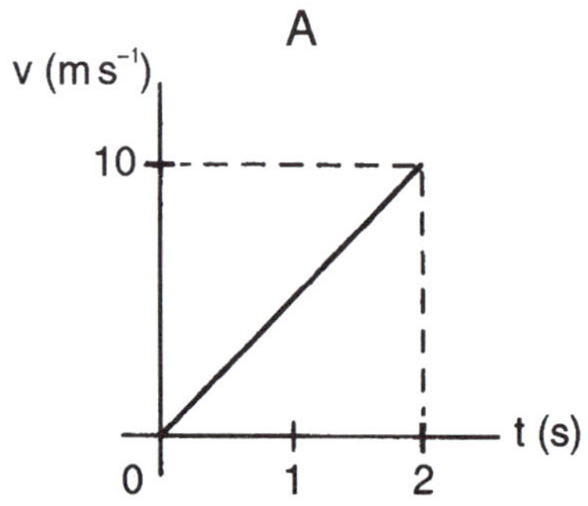

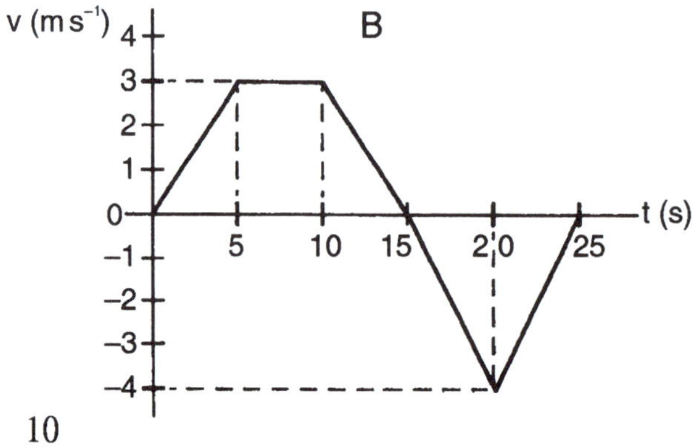

EQUATIONS OF MOTION AND PROJECTILES

Memorise

1. The equations:- $v = u + at$ $\quad$ $s = ut + \frac{1}{2}at^2$ $\quad$ $v^2 = u^2 + 2as$

and (sorry about this), you should also be able to derive the three equations as shown below:-

2. By definition, the acceleration is given by $a = \dfrac{v - u}{t}$

$$\Rightarrow \quad v - u = at$$

$$\Rightarrow \quad \boxed{v = u + at}$$

3. s = area under v/t graph

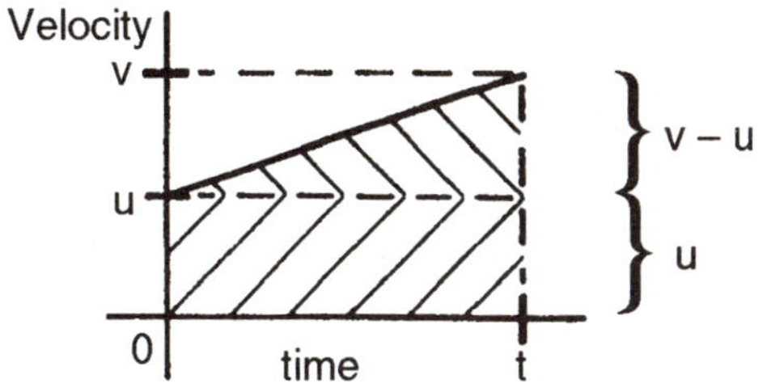

s =  + 

$$= ut + \tfrac{1}{2}(v - u)t$$

$$= ut + \tfrac{1}{2}(at)t$$

$$\Rightarrow \quad \boxed{s = ut + \tfrac{1}{2}at^2}$$

4. $v = u + at \quad \Rightarrow \quad v^2 = (u + at)^2$

$$= u^2 + 2\,uat + a^2t^2$$

$$= u^2 + 2a\,(ut + \tfrac{1}{2}at^2) \quad \Rightarrow \quad \boxed{v^2 = u^2 + 2\,as}$$

Understand

5. A **method** for tackling equations of motion problems:-

Write down all the **symbols** like this: u = $\quad$ s = $\quad$ v = $\quad$ t = $\quad$ a =

Look for **numbers** in the question, e.g. "a car accelerates at $2\,m\,s^{-2}$".
Fill in $\quad a = 2\,m\,s^{-2}$

Look for **hidden values** e.g. the words "from rest".
Fill in $\quad u = 0\,m\,s^{-1}$

If two **directions** are involved use this "direction diagram" to decide whether a number is positive or negative.

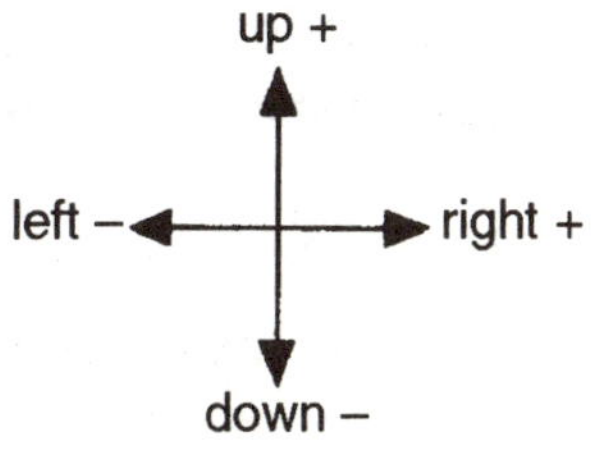

Try to choose the best equation for the job.

Solve

6. An object is fired vertically upwards at $50\,\text{m s}^{-1}$. Assuming $g = 10\,\text{m s}^{-2}$ and air resistance is negligible, (a) how long does it take to reach its maximum height?
(b) find the maximum height.
(c) how fast is it travelling at a point 80 m above the ground?

Remember the method?

Symbols:
$u =$
$s =$
$v =$
$t =$
$a =$

Numbers: $u = 50\,\text{m s}^{-1}$

Hidden values: a = acceleration due to gravity
$v = 0$ (at the top)

Directions: a is negative (down) u is positive (up)

ANSWERS

$u = 50\,\text{m s}^{-1}$
$v = 0$
$a = -10\,\text{m s}^{-2}$
$s =$
$t =$

(a) $v = u + at \Rightarrow 0 = 50 - 10t$
$\Rightarrow$ $\mathbf{t = 5\,s}$

(b) $s = ut + \frac{1}{2}at^2 \Rightarrow s = (50 \times 5) + (\frac{1}{2} \times -10 \times 25)$
$\Rightarrow s = 250 - 125$
$\Rightarrow$ $\mathbf{s = 125\,m}$

(c) $v^2 = u^2 + 2as \Rightarrow v^2 = 50^2 + (2 \times -10 \times 80)$
$\Rightarrow v^2 = 2500 - 1600$
$\Rightarrow v^2 = 900$
$\Rightarrow$ $\mathbf{v = 30\,m\,s^{-1}}$

Understand

7. The motion of a **projectile** consists of two independent parts or components:
(a) constant horizontal velocity (b) constant vertical acceleration

First look at the example over the page with ball A and ball B, then notice how similar this is to the projectile problem in the **Solve** section.

Example

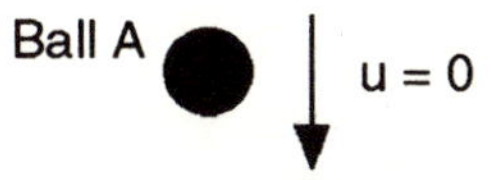
Ball A
u = 0
Ball B
5 m s⁻¹

Ball A falls from rest 45 m to the ground. Ball B rolls along a frictionless horizontal surface at $5\,m\,s^{-1}$. Find:-

(a) the time taken for A to reach the ground

(b) the distance travelled by B in that time

(c) the velocity of A just before impact

(d) the velocity of B at that time.

ANSWERS

Vertical ball A

(a) $s = ut + \frac{1}{2}at^2$

$\Rightarrow -45 = 0 - 5t^2$

$\Rightarrow$ **$t = 3\,s$**

(c) $v = u + at$

$= 0 - 10 \times 3$

$\Rightarrow$ **$v = -30\,m\,s^{-1}$**

Horizontal ball B

(b) $s = vt$

$= 5 \times 3$

$\Rightarrow$ **$s = 15\,m$**

(d) **$v = 5\,m\,s^{-1}$**

(constant velocity)

Solve

8. A projectile is fired horizontally off a cliff as shown:

Find:- (a) the time of flight

(b) the range

(c) the vertical velocity

(d) the horizontal velocity

(e) the actual (resultant) velocity

} just before impact.

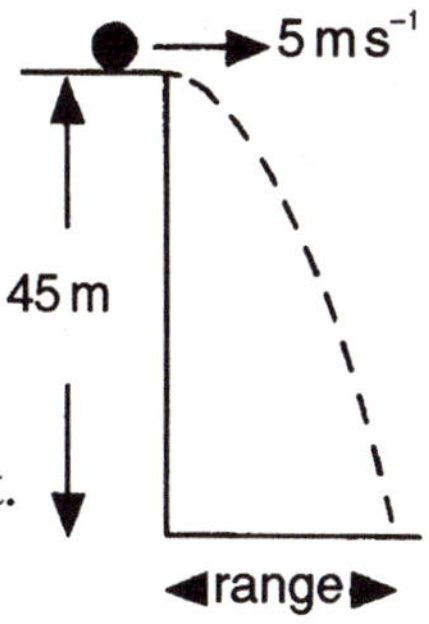
5 m s⁻¹
45 m
range

ANSWERS

Vertical Motion

(a) $s = ut + \frac{1}{2}at^2$

$\Rightarrow -45 = 0 - 5t^2$

$\Rightarrow$ **$t = 3\,s$**

(c) vertical velocity

$v = u + at$

$= 0 - 10 \times 3$

$\Rightarrow$ **$v = -30\,m\,s^{-1}$**

Horizontal Motion

(b) Range $s = vt = 5 \times 3$

$\Rightarrow$ **$s = 15\,m$**

(d) Horizontal Velocity **$v = 5\,m\,s^{-1}$**

(e)

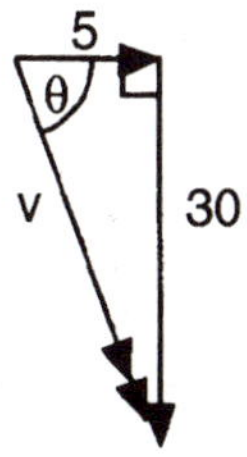
5
θ
v
30

(from diagram) $v^2 = 5^2 + 30^2$

$\Rightarrow v = 30.4\,m\,s^{-1}$

$\tan\theta = \frac{30}{5} \Rightarrow \theta = 80.5°$

Actual velocity = **$30.4\,m\,s^{-1}$,**
80.5° below the horizontal

Solve

9. A is projected up from the ground
B travels at constant velocity as shown

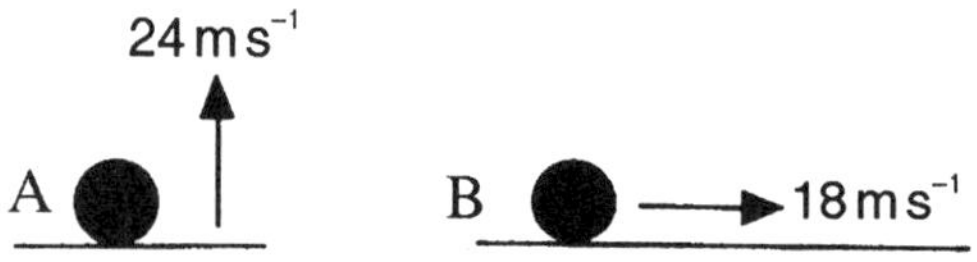

Find:- (a) the maximum height reached by A
(b) the time for A to reach maximum height
(c) the total flight time for A
(d) the distance gone by B during A's up and down journey.

ANSWERS

Vertical ball A

(a) $v^2 = u^2 + 2as$

$\Rightarrow\ 0 = 24^2 - 20s \Rightarrow$ **s = 28.8 m**

(b) $v = u + at$

$\Rightarrow\ 0 = 24 - 10t \Rightarrow$ **t = 2.4 s**

(c) Total flight time (up and down)
$= 2 \times 2.4 =$ **4.8 s**

Horizontal ball B

(d) $s = vt$
$= 18 \times 4.8$
= 86.4 m

10. A projectile is launched with the initial velocity shown.
Find:- (a) the maximum height
(b) the total flight time
(c) the (horizontal) range.

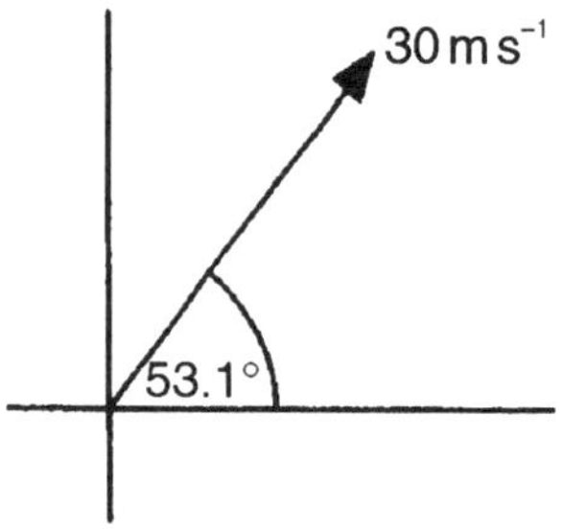

Hint:
Remember page 7, item 5.

ANSWERS

Vertical Motion

(a) $v^2 = u^2 + 2as$

$\Rightarrow\ 0 = (30 \sin 53.1°)^2 - 20s$

$\Rightarrow\ 0 = (30 \times 0.8)^2 - 20s$

$\Rightarrow$ **s = 28.8 m**

(b) $v = u + at$

$\Rightarrow\ 0 = (30 \sin 53.1°) - 10t$

$\Rightarrow\ 0 = (30 \times 0.8) - 10t$

$\Rightarrow$ **t = 2.4 s**

$\Rightarrow$ Total flight time = **4.8 s**

Horizontal Motion

(c) $s = vt$
$= (30 \cos 53.1°) \times 4.8$
$= (30 \times 0.6) \times 4.8$
= 86.4 m

FORCES

Memorise

1. $F = ma$ $F = \dfrac{mv - mu}{t}$ Units of force: N

2. One newton is the (unbalanced) force which gives a 1 kg mass an acceleration of $1\,m\,s^{-2}$.

3. Force is a vector.

Understand

4. An accelerating object has an unbalanced (resultant) force, F, on it acting in the same direction as the acceleration.

5. If an object is not accelerating then the unbalanced force $F = 0$.

6. Some tips for solving force problems:-
(a) Draw a **free body diagram** showing the forces acting on the object and the direction of its acceleration.
(b) Try to deal with **one object** at a time, but remember one useful exception: when objects are tied together, they behave like a single larger object.
(c) Use $\mathbf{F = ma}$ or $\mathbf{a = \dfrac{F}{m}}$ as required, but you must understand that **F** is the **unbalanced force**. Now try these:-

Solve

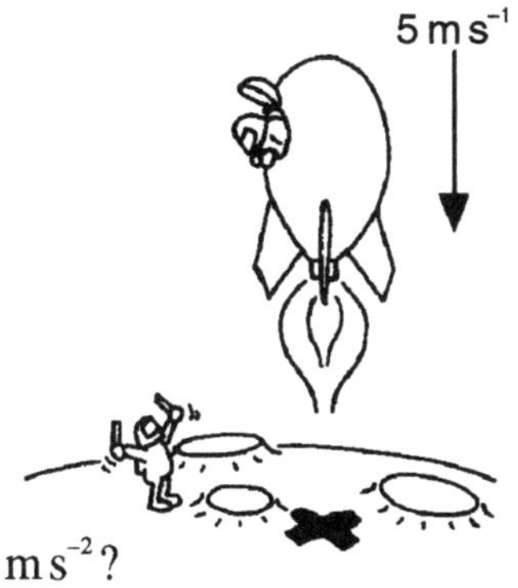

7. A 3000 kg spacecraft is descending at a constant speed of $5\,m\,s^{-1}$ on to the Moon (where $g = 1.7\,m\,s^{-2}$).
(a) Find the thrust of its rocket motors.
(b) What thrust would be needed to accelerate the rocket up off the Moon at $12\,m\,s^{-2}$?
(c) What thrust would produce a downward acceleration of $1\,m\,s^{-2}$?

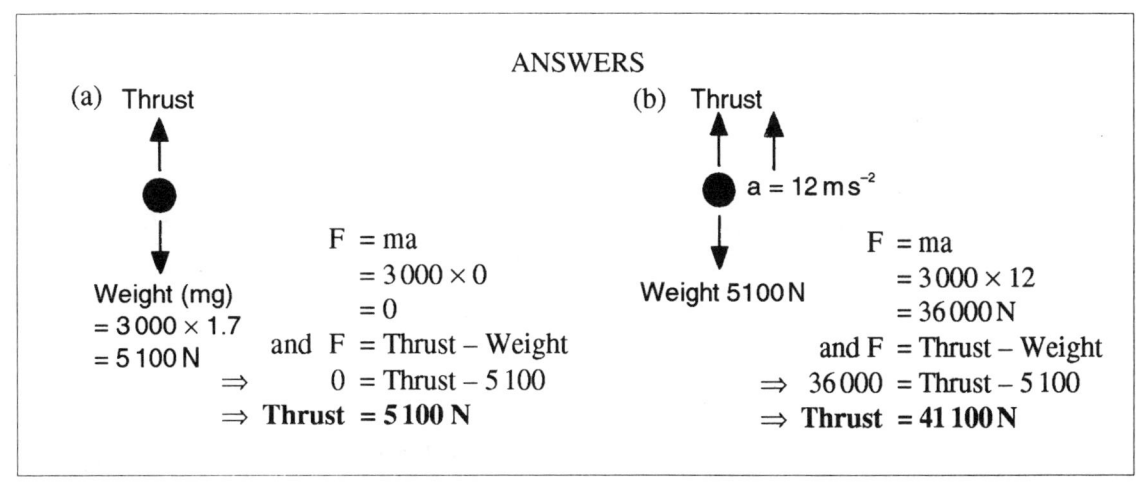

ANSWERS

(a)

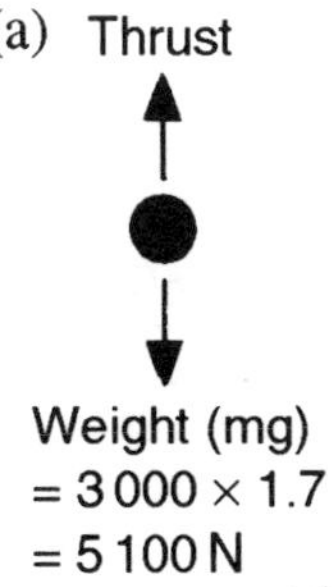

$F = ma$
$= 3000 \times 0$
$= 0$
and F = Thrust – Weight
$\Rightarrow$ 0 = Thrust – 5 100
$\Rightarrow$ **Thrust = 5 100 N**

(b)

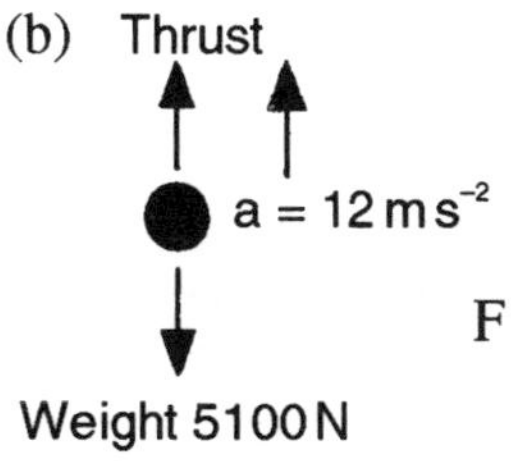

$F = ma$
$= 3000 \times 12$
$= 36\,000\,N$
and F = Thrust – Weight
$\Rightarrow$ $36\,000$ = Thrust – 5 100
$\Rightarrow$ **Thrust = 41 100 N**

ANSWERS (cont.)

(c)

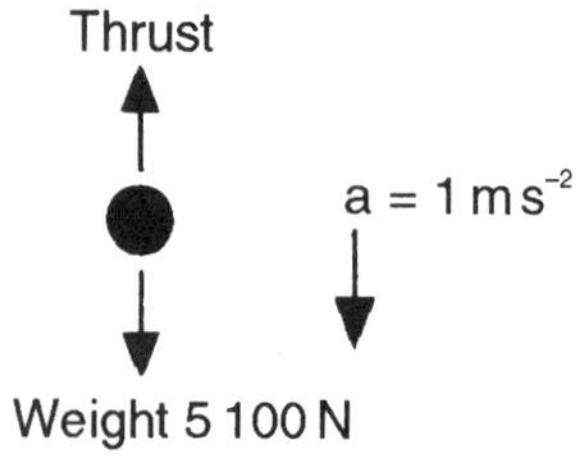
Thrust
a = 1 ms^{-2}
Weight 5 100 N

$$F = ma = 3000 \times 1 = 3000\,N$$

$$F = \text{Weight} - \text{Thrust}$$

$\Rightarrow$ $3000 = 5100 - \text{Thrust}$

$\Rightarrow$ **Thrust = 2100 N**

8. Find (a) the acceleration of the trolleys on the horizontal frictionless surface
(b) the tension T_2
(c) the tension T_1.

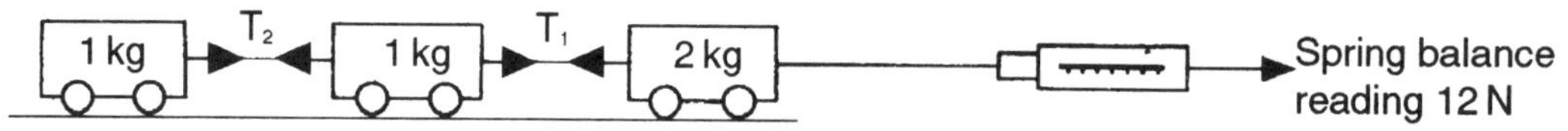
1 kg
T2
1 kg
T1
2 kg
Spring balance
reading 12 N

ANSWERS

(a) They are tied together so:-

4 kg
12 N

$$a = \frac{F}{m} = \frac{12}{4}$$

$$\mathbf{a = 3\,ms^{-2}}$$

(b)

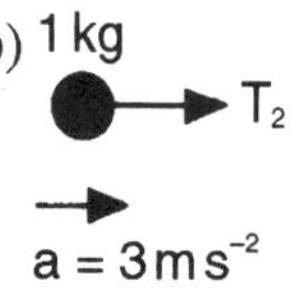
1 kg
T2
a = 3 ms^{-2}

$F = ma$

$\Rightarrow T_2 = ma = 1 \times 3$

$\Rightarrow \mathbf{T_2 = 3\,N}$

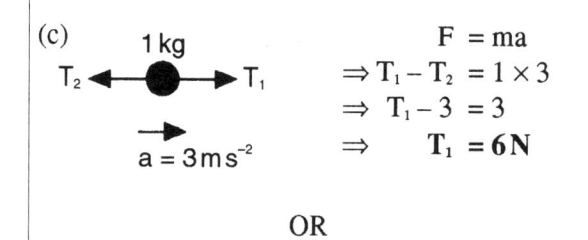

(c)

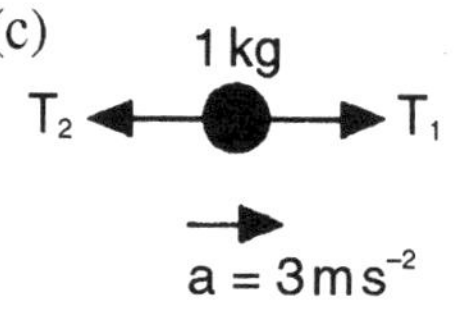
1 kg
T2
T1
a = 3 ms^{-2}

$F = ma$

$\Rightarrow T_1 - T_2 = 1 \times 3$

$\Rightarrow T_1 - 3 = 3$

$\Rightarrow \mathbf{T_1 = 6\,N}$

OR

(c)

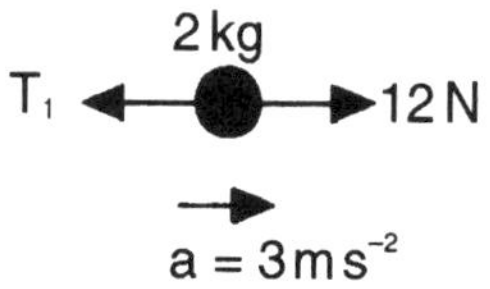
2 kg
T1
12 N
a = 3 ms^{-2}

$F = ma$

$\Rightarrow 12 - T_1 = 2 \times 3$

$\Rightarrow 12 - T_1 = 6$

$\Rightarrow \mathbf{T_1 = 6\,N}$

9. Find the acceleration of the 10 kg mass.

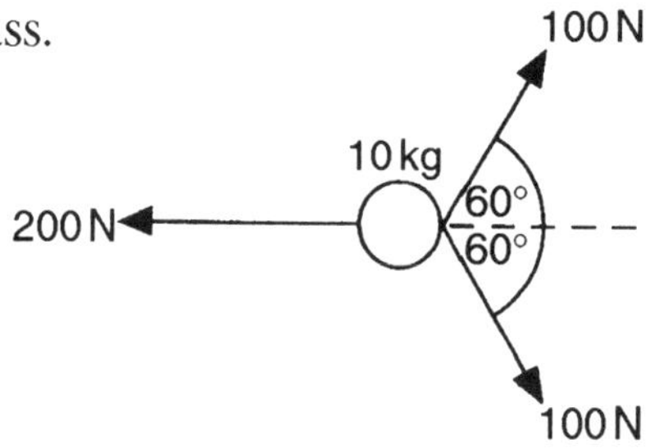
100 N
10 kg
60°
60°
200 N
100 N

ANSWER

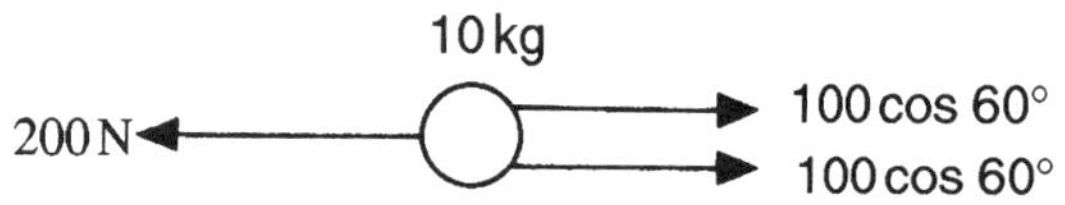
10 kg
200 N
100 cos 60°
100 cos 60°

$$a = \frac{F}{m} = \frac{2(100\cos 60) - 200}{10} = \frac{-100}{10} = -10\,ms^{-2}$$

$\Rightarrow$ **acceleration = 10 ms^{-2} to the left**

COLLISIONS AND MOMENTUM

Memorise

1. Momentum = Mass × Velocity $p = mv$ Units of momentum: $kg\,m\,s^{-1}$

2. Momentum is a vector.

State

3. The Law of Conservation of Momentum:- In any collision or explosion free of external forces, the total momentum remains the same.

4. Some details of collisions and explosions:-

Event	*Momentum (p)*	*Kinetic Energy (E_k)*
Inelastic Collision	conserved	reduces
Elastic Collision	conserved	conserved
Explosion	conserved	increases

Solve

5. Find the velocity of the trolleys when they stick together after colliding. Show that the collision is inelastic.

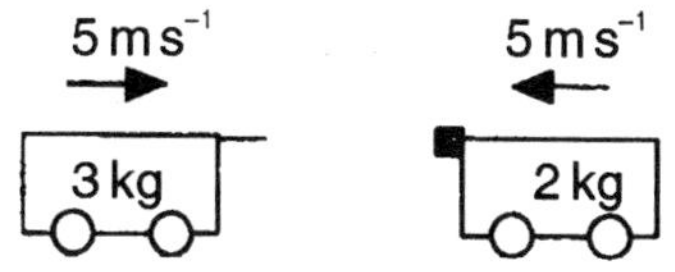

ANSWERS

Momentum before collision = Momentum after collision

$(3 \times 5) + (2 \times -5) = (3 + 2)v \Rightarrow$ **$v = 1\,m\,s^{-1}$** (to the right)

E_k before $= (\frac{1}{2} \times 3 \times 5^2) + (\frac{1}{2} \times 2 \times 5^2) =$ **62.5 J**

E_k after $= (\frac{1}{2} \times 5 \times 1^2) =$ **2.5 J**

} Remember, E_k is a scalar so no minuses

The E_k reduces (from 62.5 J to 2.5 J), so the collision is **inelastic**.

6. Find the velocity of the 10 kg ball after the collision then show that the collision is elastic.

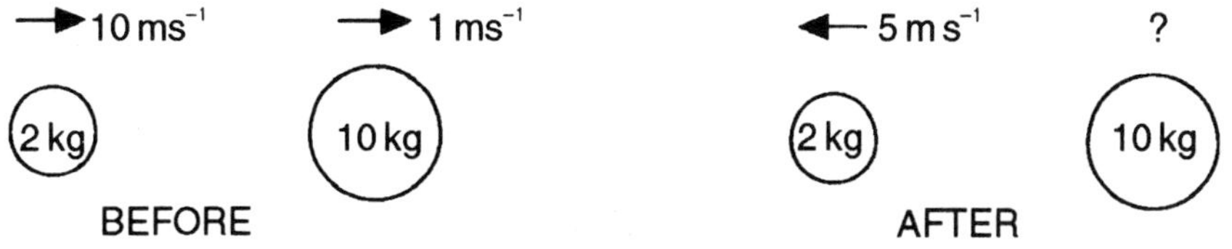

ANSWER

Momentum before collision = Momentum after collision

$(2 \times 10) + (10 \times 1) = (2 \times -5) + 10v$

$30 = -10 + 10v \Rightarrow$ **$v = 4\,m\,s^{-1}$** (to the right)

ANSWER (cont.)

E_k before $= (\frac{1}{2} \times 2 \times 10^2) + (\frac{1}{2} \times 10 \times 1^2)$
$= 105\,J$

E_k after $= (\frac{1}{2} \times 2 \times 5^2) + (\frac{1}{2} \times 10 \times 4^2)$
$= 105\,J$

E_k is conserved so the collision is elastic.

7. Find the velocity of the 2 kg fragment after the explosion.

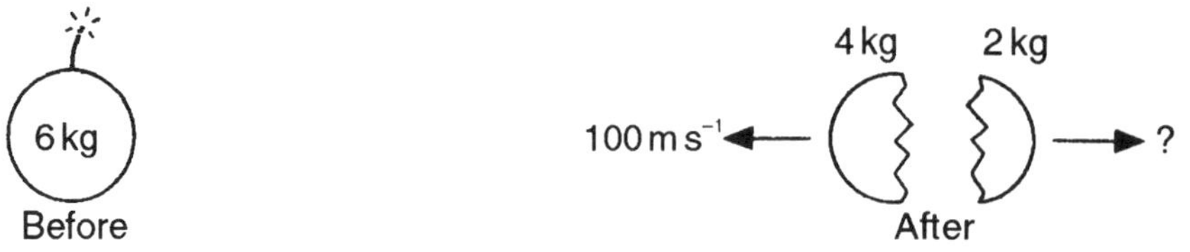

ANSWER

Momentum before = Momentum after

$0 = 4 \times (-100) + 2v$

$\Rightarrow \quad 2v = 400$

$\Rightarrow \quad$ **$v = 200\,m\,s^{-1}$ (to the right)**

FORCES AND MOMENTUM

Memorise

1. Impulse = change in momentum $\quad p = mv - mu \quad$ Units of impulse: $kg\,m\,s^{-1}$ or Ns

2. Impulse = Force × Time $\quad p = Ft$

3. Impulse is a vector.

Understand

4. When two objects interact, their changes in momentum (mv – mu) are equal in size but opposite in direction. To check that this is true, use the figures from the collision on the previous page.

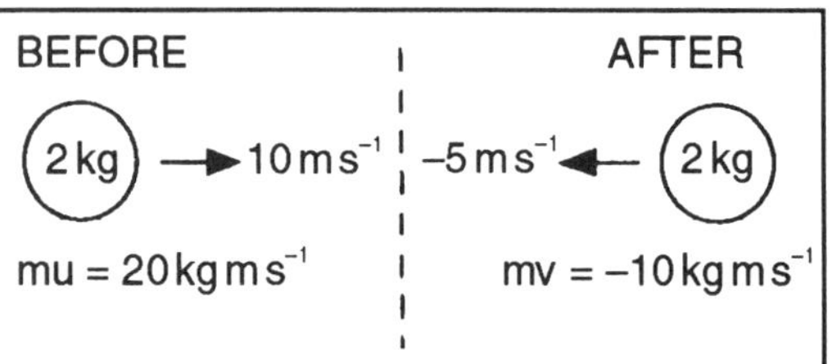

$(mv - mu) = $ **$-30\,kg\,m\,s^{-1}$**

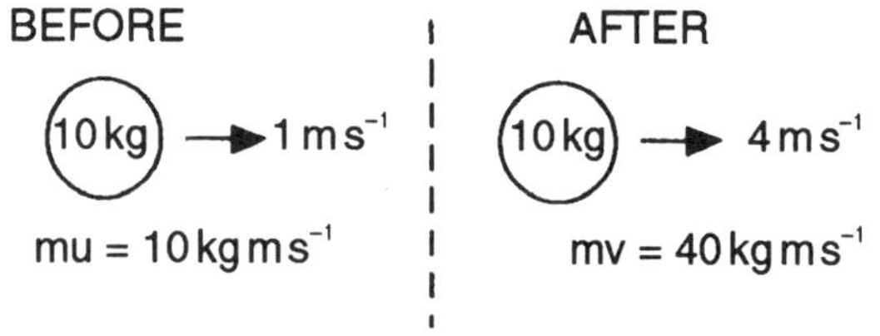

$(mv - mu) = $ **$+30\,kg\,m\,s^{-1}$**

5. If the collision lasted 0.1 s then $F = \dfrac{mv - mu}{t}$

$\Rightarrow$ Force on 2 kg ball $= \dfrac{-30}{0.1} = $ **$-300\,N$** $\qquad$ Force on 10 kg ball $= \dfrac{+30}{0.1} = $ **$+300\,N$**

When two objects interact, the forces they exert on each other are equal in size but opposite in direction. This is known as **Newton's Third Law**.

Solve

6. A 50 g golf ball is hit by a club and moves off at $18\,m\,s^{-1}$.
The ball is in contact with the club head for a time of 6 ms.
Find:- (a) the change of momentum of the ball
(b) the impulse on the ball
(c) the force of the club head on the ball
(d) the force of the ball on the club head.

ANSWERS

(a) $mv - mu = 0.05 \times 18 - 0 = \mathbf{0.9\,kg\,m\,s^{-1}}$

(b) impulse, p = change of momentum
so answer is same as for (a) $\mathbf{0.9\,kg\,m\,s^{-1}}$
or **0.9 Ns**

(c) $F = \frac{p}{t} = \frac{0.9}{0.006} = \mathbf{150\,N}$

(d) By Newton's 3rd Law, $F = \mathbf{-150\,N}$

DENSITY

Memorise

1. $\text{Density} = \frac{\text{Mass}}{\text{Volume}}$ $\qquad \rho = \frac{m}{V}$ $\qquad$ Units of density: $kg\,m^{-3}$

Describe

2. An experiment to measure the density of air:-

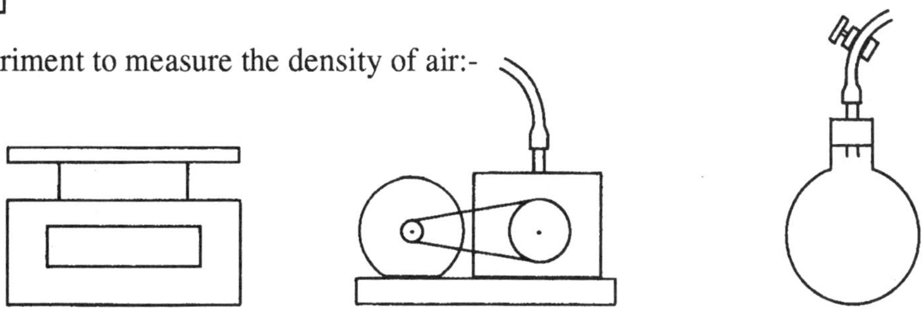

sensitive electronic balance — powerful vacuum pump — round bottom glass flask

You can obtain a reasonable value for the density of air using the above equipment. Make a brief note now of the measurements needed and what you would do.

Understand

3. Solid and liquid densities are of a similar size because their molecular spacing is similar. The spacing between gas molecules is about ten times greater, so the same mass (number of molecules) of a gas will occupy a volume $10 \times 10 \times 10 = 1000$ times greater, making gas densities 1000 times smaller.

PRESSURE AND THE GAS LAWS

Memorise

1. Pressure = $\frac{\text{Force}}{\text{Area}}$ $\boxed{P = \frac{F}{A}}$ Units of pressure: pascals, Pa

2. 1 pascal = 1 newton per square metre $1\ Pa = 1\ Nm^{-2}$

Solve

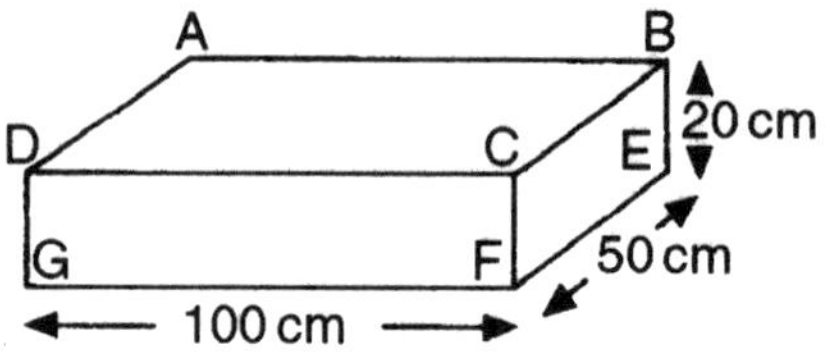

3. Find the pressure exerted by this 12 kg block on the floor:-
   (a) when it is resting on side ABCD (b) when it is resting on side BCFE

ANSWERS

(a) $P = \frac{F}{A} = \frac{mg}{A} = \frac{12 \times 10}{1 \times 0.5} = \frac{120}{0.5} = \mathbf{240\ Pa}$ (b) $P = \frac{F}{A} = \frac{mg}{A} = \frac{12 \times 10}{0.2 \times 0.5} = \frac{120}{0.1} = \mathbf{1\,200\ Pa}$

State

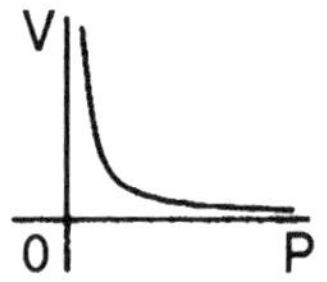

4. **Boyle's Law**
   The volume of a fixed mass of gas varies inversely with pressure (provided its temperature doesn't change).

5. **The Pressure Law**
   The pressure of a fixed mass of gas varies directly with Kelvin temperature (provided the volume doesn't change).

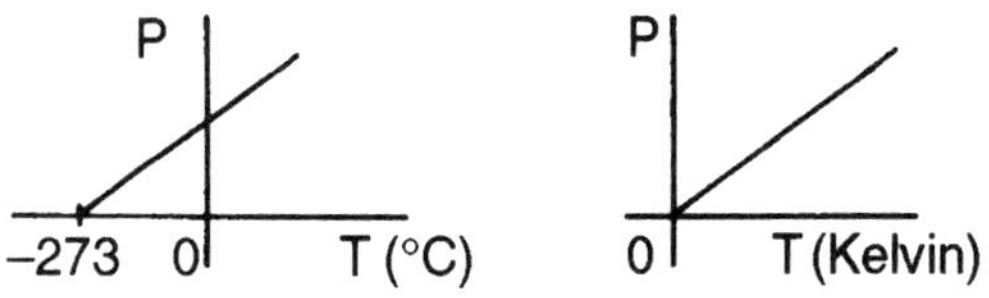

Memorise

6. $\boxed{P_1V_1 = P_2V_2}$

7. $\boxed{\frac{P_1}{T_1} = \frac{P_2}{T_2}}$ (T in Kelvin)

8. Kelvin temp. = Celsius temp. + 273

9. Celsius temp. = Kelvin temp. – 273

Solve

10. A steel cylinder contains oxygen at a pressure of 2×10^7 Pa and a temperature of 27°C. The temperature is raised to 57°C. Find the new pressure.

ANSWER If volume and mass don't change,

$$\frac{P_1}{T_1} = \frac{P_2}{T_2} \Rightarrow \frac{2 \times 10^7}{300} = \frac{P_2}{330} \Rightarrow P_2 = \frac{330 \times 2 \times 10^7}{300} = \mathbf{2.2 \times 10^7\ Pa}$$

11. Atmospheric pressure is equivalent to the pressure of 10 m of water. An air bubble has a volume of $4 cm^3$ at a depth of 30 m in a pond. What would be its volume just as it reaches the surface?

> ANSWER
>
> Assuming mass and temperature don't change, $P_1 V_1 = P_2 V_2$
>
> $P_1 = 10 + 30 = 40 \quad P_2 = 10 \quad P_1 V_1 = P_2 V_2 \Rightarrow 40 \times 4 = 10 \times V_2 \Rightarrow$ $\mathbf{V_2 = 16 cm^3}$

Understand

12. **Gas pressure** is the result of millions of gas molecules colliding with a surface and exerting force on a given area.

13. **Temperature** of a gas depends on the average kinetic energy, E_k of the molecules.

14. **Boyle's Law**
If volume decreases, the molecules have a smaller distance to travel between collisions with the walls of the container so you get more collisions per second. This means more force. There is also a smaller area. Both these effects mean that pressure increases.

15. **The Pressure Law**
If temperature increases, E_k increases so molecular speed increases. This causes harder collisions and also more collisions per second. Both these effects mean that pressure increases.

Memorise

16. Pressure due to a liquid varies directly with depth, h and density, ρ. $P = \rho g h$

Understand

17. When an object is immersed in a liquid, there is an upthrust (buoyancy force) on the object.

18. The buoyancy force equals the weight of liquid displaced.

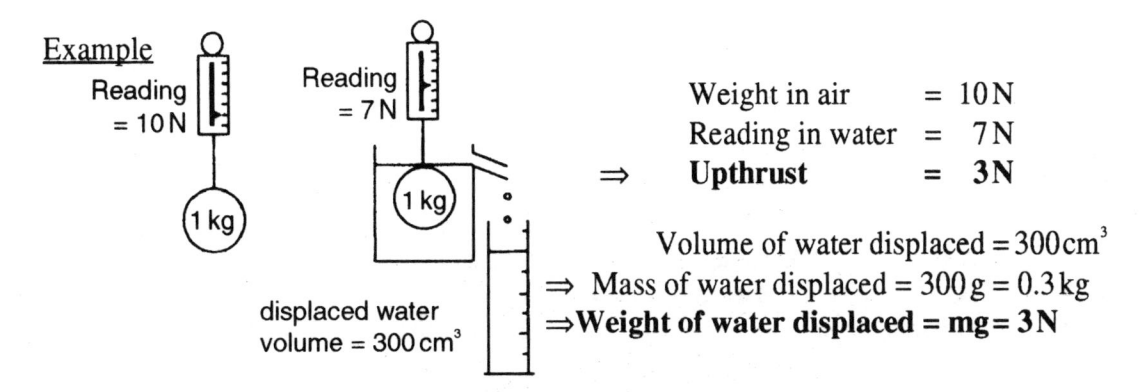

Example

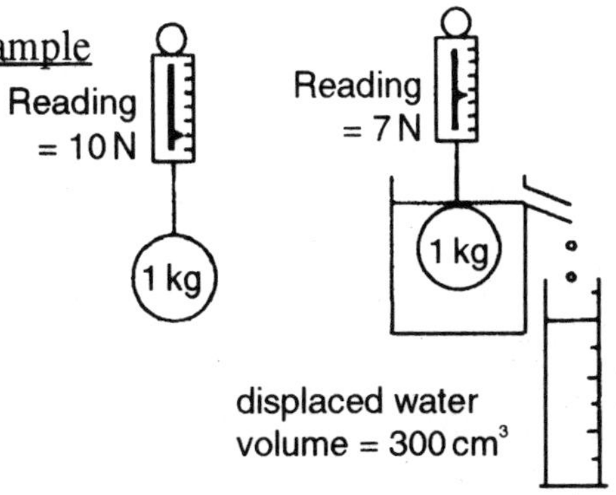

Weight in air	=	10 N
Reading in water	=	7 N
⇒ **Upthrust**	=	**3 N**

Volume of water displaced = $300 cm^3$
⇒ Mass of water displaced = 300 g = 0.3 kg
⇒**Weight of water displaced = mg = 3 N**

19. If upthrust = weight of object it floats.
If upthrust < weight of object it sinks.
If upthrust > weight of object it rises.

3. ELECTRICITY AND ELECTRONICS

ELECTRIC FIELDS AND POTENTIAL DIFFERENCE

State

1. In an **electric field**, a **charge** experiences a **force**.

Understand

2. Electric field lines show the direction of the force (on a positive charge).

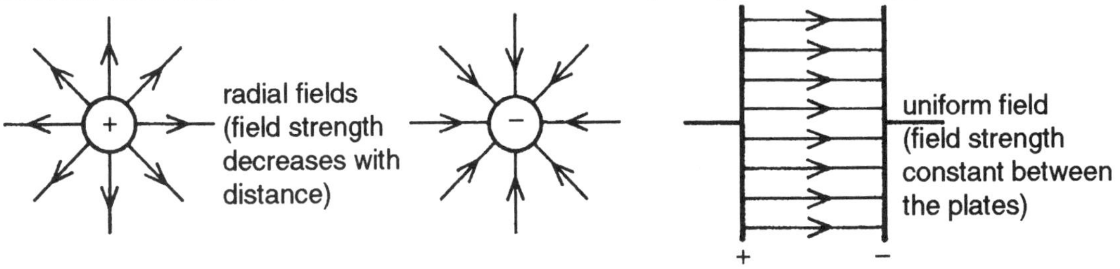

3. If a **charge** is **moved** in a field then **work** is **done**.

Memorise

4. Potential difference $= \dfrac{\text{Work done}}{\text{charge}}$ $\boxed{V = \dfrac{W}{Q}}$ Units of p.d.: volts, V

5. 1 volt = 1 joule per coulomb $1\,V = 1\,JC^{-1}$

Solve

6.

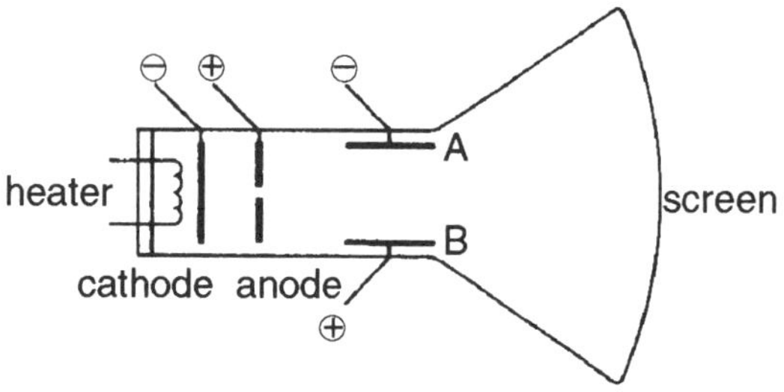

Potential difference across cathode/anode $= 4\,kV$
Electron charge $= 1.6 \times 10^{-19}\,C$
Electron mass $= 9 \times 10^{-31}\,kg$

(a) Describe the motion of an electron:-
 (i) between cathode and anode
 (ii) after passing through hole in anode
 (iii) as it passes between plates AB.

(b) Find:-
 (i) the work done on an electron between cathode and anode
 (ii) kinetic energy gained by the electron
 (iii) speed of the electron on reaching anode.

ANSWERS

(a) (i) acceleration due to the force created by the electric field
 (ii) constant speed in a straight line (no field; no force)
 (iii) parabolic motion (this is similar to projectile motion)

(b) (i) $V = \dfrac{W}{Q} \Rightarrow W = QV = 1.6 \times 10^{-19} \times 4 \times 10^{3} = \mathbf{6.4 \times 10^{-16}\,J}$
 (ii) the kinetic energy gained = the work done by the field = $\mathbf{6.4 \times 10^{-16}\,J}$

ANSWERS (cont.) (iii) if the electron started from rest then

$$\tfrac{1}{2}mv^2 = 6.4 \times 10^{-16} \quad \Rightarrow \quad v^2 = \frac{2 \times 6.4 \times 10^{-16}}{9 \times 10^{-31}} \quad \Rightarrow \quad \mathbf{v = 3.8 \times 10^7\,m\,s^{-1}}$$

V, I AND R

Understand

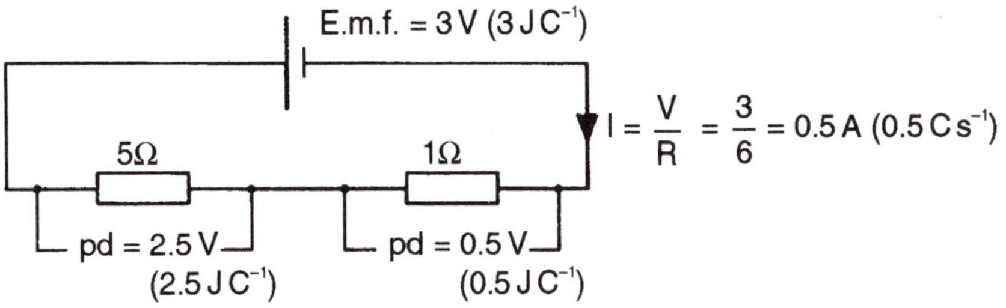

1. **The e.m.f.** (electromotive force) is the **energy** supplied by the cell to each coulomb of **charge**.

2. **The p.d.** (potential difference) is the **work** required to push each coulomb of **charge** through a resistor. In resistors the work done becomes heat.

3. If energy is conserved in a series circuit then e.m.f. = sum of all the potential differences.
   In the circuit above:- $3\,V = 2.5\,V + 0.5\,V$

4. **Current** is the rate of flow of charge i.e. $\dfrac{\textbf{charge}}{\textbf{time}}$, the number of coulombs passing per second.

Memorise

5. $V = IR$ $\quad V = \dfrac{W}{Q}$ $\quad I = \dfrac{Q}{t}$ $\quad \text{Energy} = QV = ItV$

$$\text{Power} = \frac{\text{Energy}}{\text{time}} = \frac{ItV}{t} = IV$$

6. Resistors in series:-

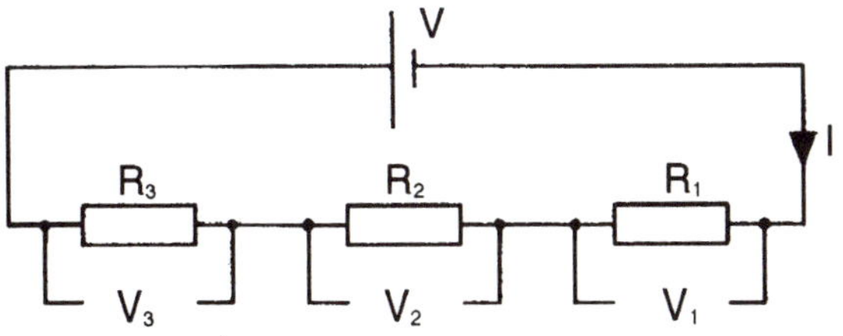

$V = V_1 + V_2 + V_3$

I is same in each series resistor.

If R is the total resistance,

$IR = IR_1 + IR_2 + IR_3$

$\Rightarrow IR = I(R_1 + R_2 + R_3)$

$\Rightarrow \mathbf{R = R_1 + R_2 + R_3}$

7. Resistors in parallel:-

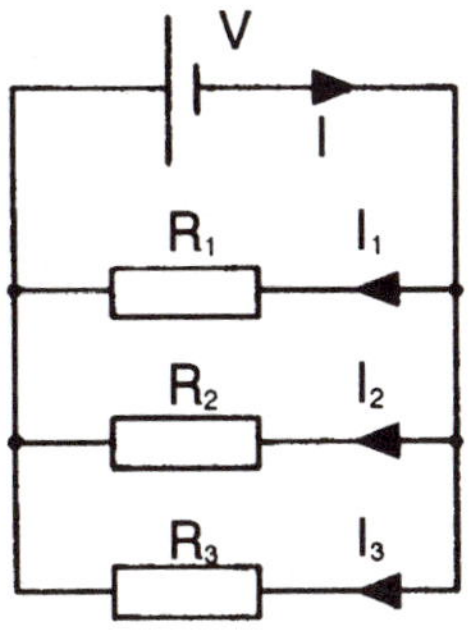

$I = I_1 + I_2 + I_3$

If R is the total resistance,

$$\frac{V}{R} = \frac{V}{R_1} + \frac{V}{R_2} + \frac{V}{R_3}$$

$$\Rightarrow V\left(\frac{1}{R}\right) = V\left(\frac{1}{R_1} + \frac{1}{R_2} + \frac{1}{R_3}\right)$$

$$\Rightarrow \mathbf{\frac{1}{R} = \frac{1}{R_1} + \frac{1}{R_2} + \frac{1}{R_3}}$$

Solve

8. Show that each resistor group A, B and C has a total resistance equal to this page number:-

A

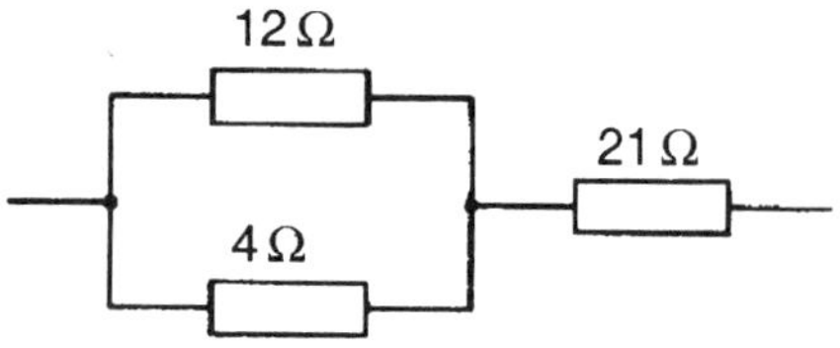

B

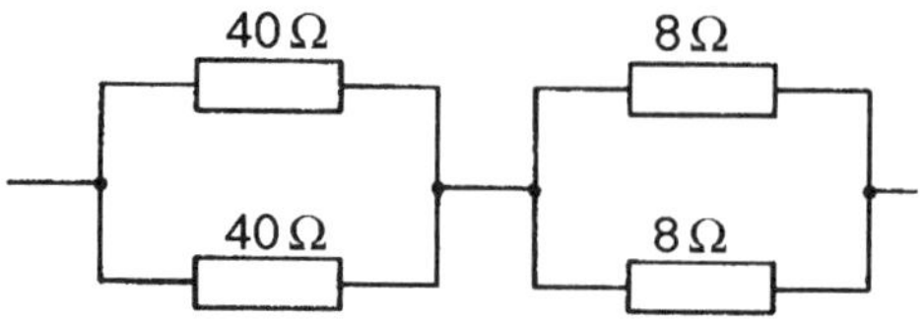

C

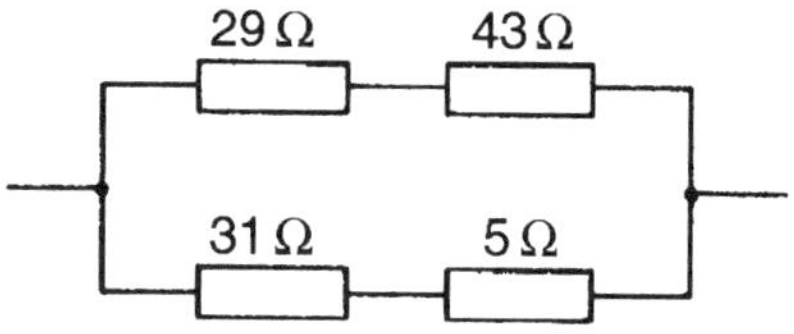

ANSWERS

(A) $\frac{1}{R} = \frac{1}{12} + \frac{1}{4} = \frac{4}{12} \Rightarrow R = \frac{12}{4} = 3\,\Omega$

$3 + 21 = \mathbf{24\,\Omega}$

(B) $\frac{1}{R} = \frac{1}{40} + \frac{1}{40} = \frac{2}{40} \Rightarrow R = 20\,\Omega$

$\frac{1}{R} = \frac{1}{8} + \frac{1}{8} = \frac{2}{8} \Rightarrow R = 4\,\Omega$

$20 + 4 = \mathbf{24\,\Omega}$

(C) $29 + 43 = 72\,\Omega$

$31 + 5 = 36\,\Omega$

$\frac{1}{R} = \frac{1}{72} + \frac{1}{36} = \frac{3}{72} \Rightarrow \mathbf{R = 24\,\Omega}$

9. Circuit A is connected to a 12 V battery.
Find:- (a) the current in the 21 Ω resistor
(b) the p.d. across it
(c) the power dissipated in the 21 Ω resistor
(d) the energy released in the 21 Ω resistor in 10 s.

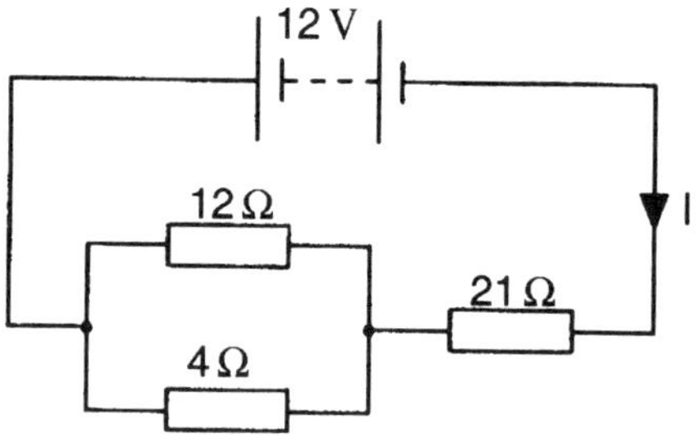

ANSWERS

(a) $I = \frac{V}{R} = \frac{12}{24} = \mathbf{0.5\,A}$

(b) $V = IR = 0.5 \times 21 = \mathbf{10.5\,V}$

(c) $P = VI = 10.5 \times 0.5 = \mathbf{5.25\,W}$

(d) $E = ItV = 0.5 \times 10 \times 10.5 = \mathbf{52.5\,J}$

INTERNAL RESISTANCE

State

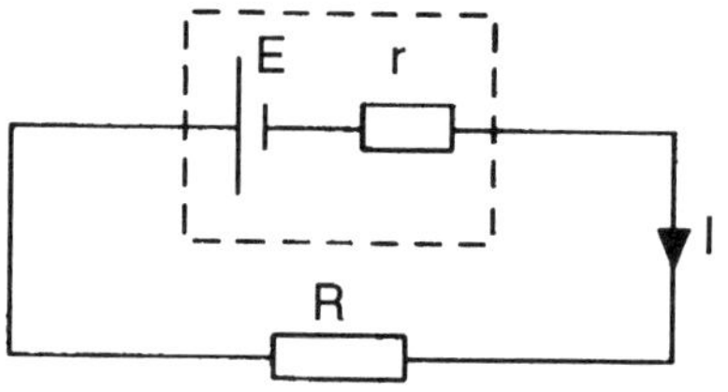

1. Real cells and batteries have internal resistance.
They are like a source of e.m.f. with a resistor in series.

Memorise

2.

E	=	V	+	v
e.m.f		terminal p.d.		"lost volts"

E	=	IR	+	Ir
e.m.f		terminal p.d.		"lost volts"

Understand

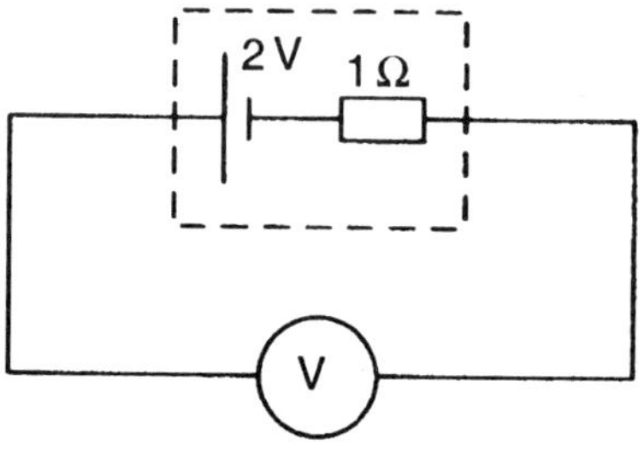

3. Voltmeters have nearly infinite resistance so $I = 0$, so $Ir = 0$ i.e. there are no lost volts.
The voltmeter reads the emf, 2 V.

4.

E =	t.p.d. + "lost volts"

In this circuit

$E = IR + Ir$
$2 = (0.5 \times 3) + (0.5 \times 1)$

The voltmeter reads the t.p.d., 1.5 V

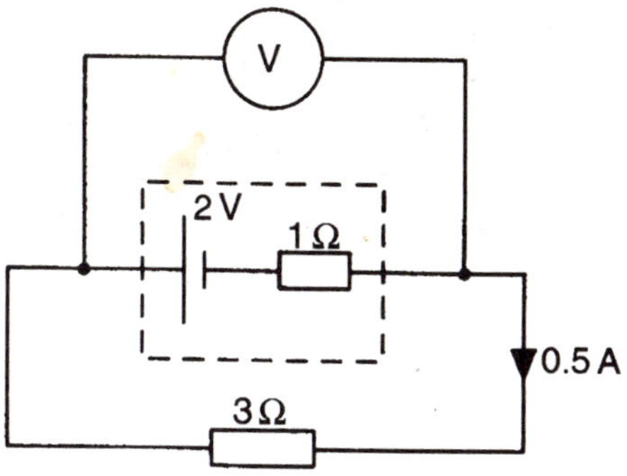

Describe

5. An experiment to measure the internal resistance of a cell.

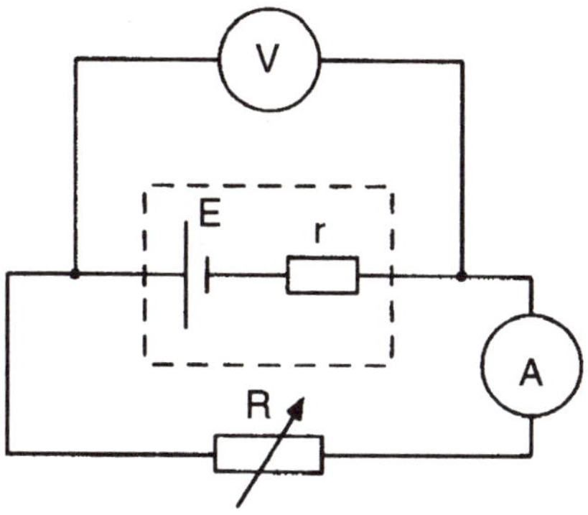

$E = IR + Ir$
$E = V + Ir$

$V = -rI + E$
$y = mx + c$

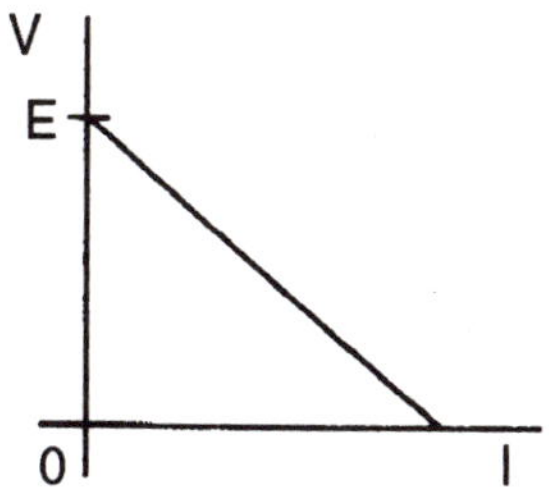

Internal resistance r = –gradient of line, m
E.m.f. E = intercept on y axis, c.

Solve

6. The heater element has a resistance of 1.5 Ω.
Its power output is 24 W. Find:-
(a) the current in the heater
(b) the voltmeter reading
(c) the internal resistance of the supply.

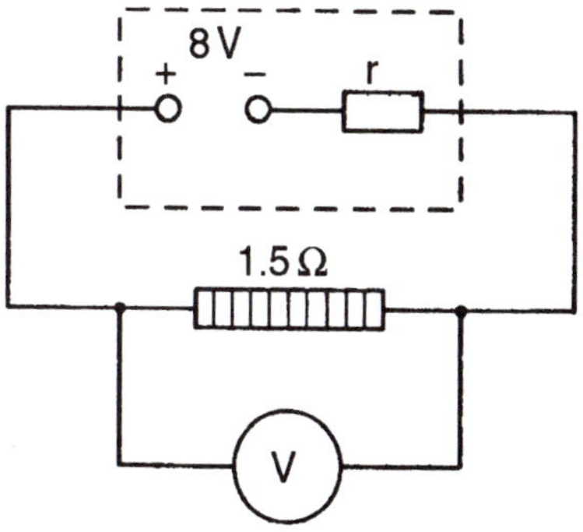

ANSWERS

(a)
$P = IV$
$= I(IR)$
$= I^2R$
$\Rightarrow 24 = I^2 \times 1.5$
$\Rightarrow I^2 = \frac{24}{1.5}$
$= 16$
$\Rightarrow$ **I = 4 A**

(b)
$V = IR$
$= 4 \times 1.5$
= **6 V** (the t.p.d.)

(c)
$E = IR + Ir$
$\Rightarrow 8 = 6 + 4r$
$\Rightarrow 4r = 2$
$\Rightarrow$ r = **0.5 Ω**

THE WHEATSTONE BRIDGE (AND METRE BRIDGE)

Memorise

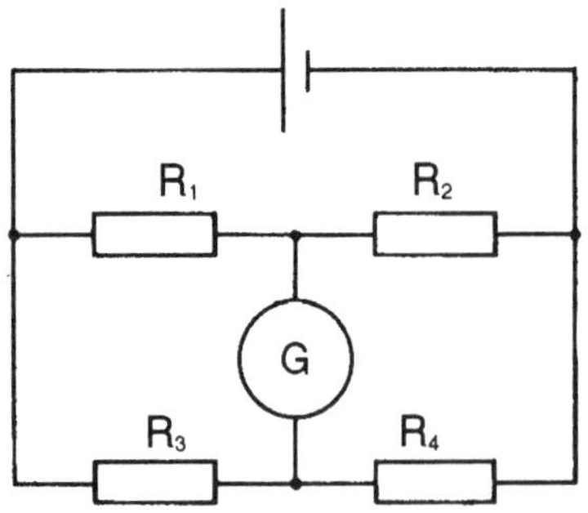

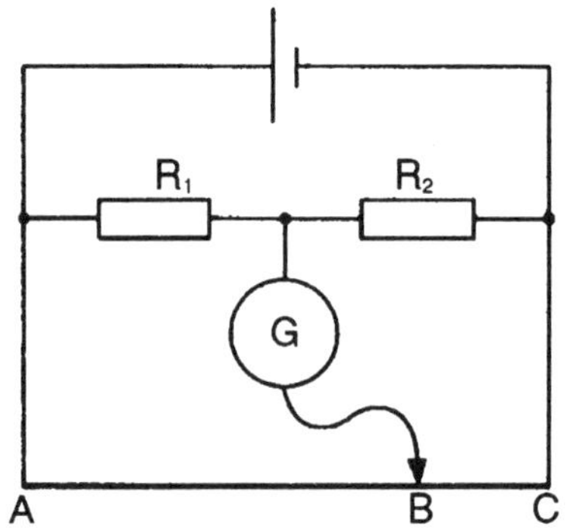

1. When the bridge is **balanced**, i.e. when the galvanometer G reads **zero**:-

$$\frac{R_1}{R_2} = \frac{R_3}{R_4} \quad \text{and} \quad \frac{R_1}{R_2} = \frac{AB}{BC}$$

Understand

2. If a bridge is initially balanced and one of the resistors is changed by a small amount ΔR, the current in the galvanometer will be proportional to the change in resistance.

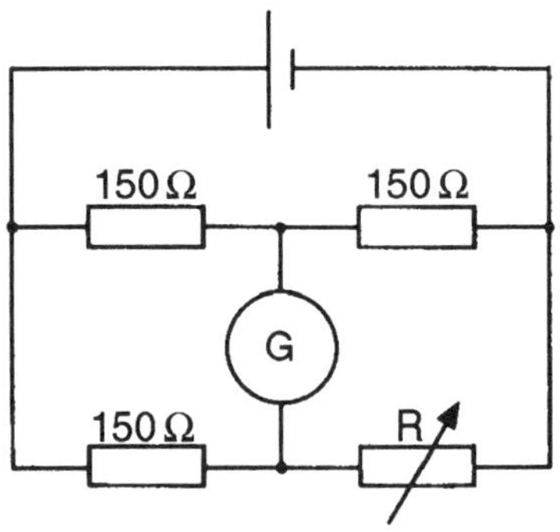

R (Ω)	ΔR (Ω)	I (mA)
150	0	0.0
151	1	0.5
152	2	1.0
153	3	1.5
154	4	2.0
155	5	2.5

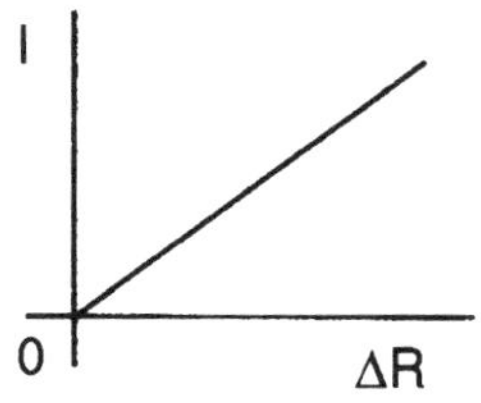

3. The variable resistor 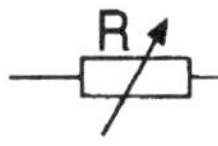 could, in practice, be a strain gauge, a thermistor or a light-dependent resistor. In this way the bridge could be used to show changes in force, temperature or light intensity.

A.C.

Memorise

1. For sinusoidal current and voltage:-

$$\text{r.m.s. value} = \frac{\text{peak value}}{\sqrt{2}}$$

Solve

2. Find peak and r.m.s. voltages. Y gain = 5 volts per cm

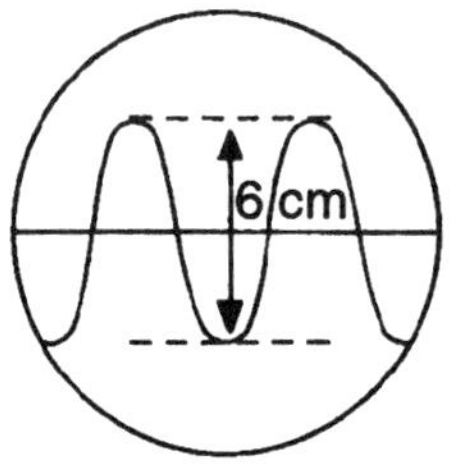

ANSWERS

Peak voltage = $3 \times 5 = \mathbf{15\,V}$ r.m.s. voltage = $\frac{15}{\sqrt{2}} = \mathbf{10.6\,V}$

Understand

3. How to measure the frequency of an a.c. source

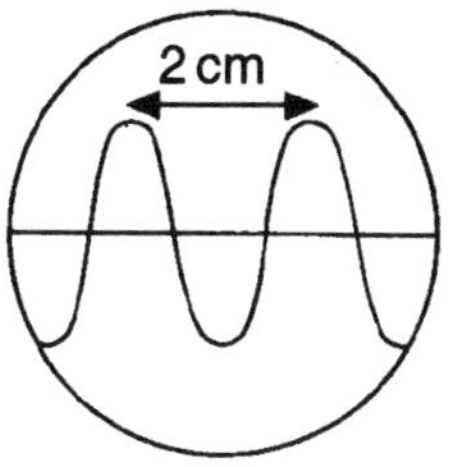

$f = \frac{1}{T}$ where T is the period, the time for 1 wave (in seconds).

$T = 20\,\text{ms} = \frac{20}{1000}\,\text{s}$ $f = \frac{1}{T} = \frac{1000}{20} = \mathbf{50\,Hz}$

Solve

4. Show that this a.c. waveform has an r.m.s. voltage of 28.3 V and a frequency of 20 kHz.

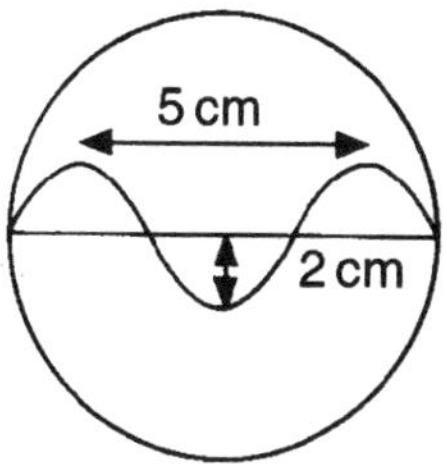

Understand

5. When a resistor is in an a.c. circuit, its resistance is unaffected by frequency. It's just the same as its resistance in a d.c. circuit.

CAPACITANCE

Memorise

1. $C = \frac{Q}{V}$ Units of capacitance: farads, F

Describe

2. How you could show that $Q \propto V$ when you charge a capacitor.

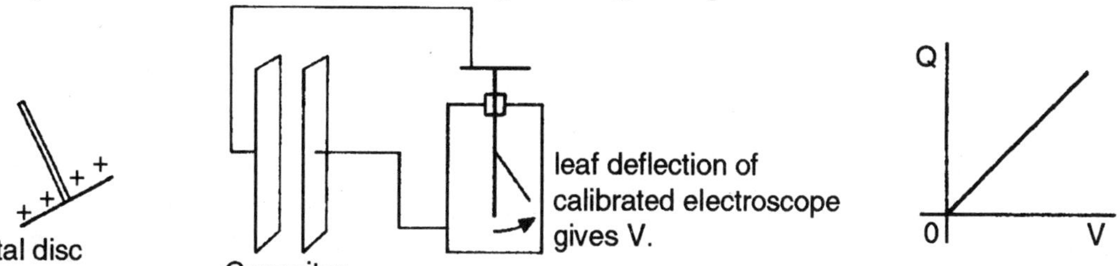

Solve

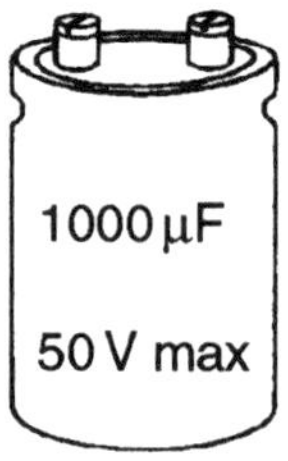

3. How much charge can this capacitor store when operating at its maximum design voltage?

ANSWER $C = \frac{Q}{V} \Rightarrow Q = CV = 1000 \times 10^{-6} \times 50$ $= \mathbf{0.05\,C}$

Understand

4. Work must be done (against the electrostatic forces) when pushing electrons on to the negative plate and pulling them off the positive plate. The work becomes energy stored in the electric field between the plates of the capacitor.

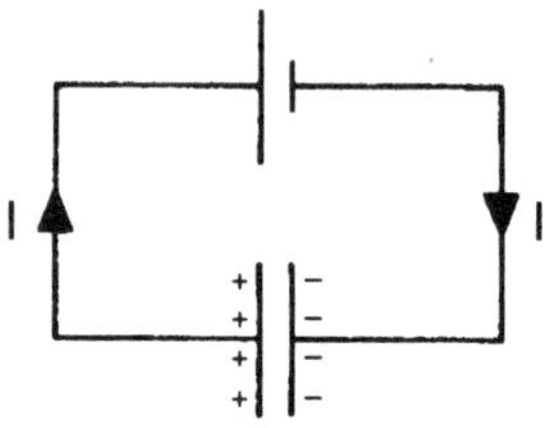

5. Energy stored = area under the Q/V graph.

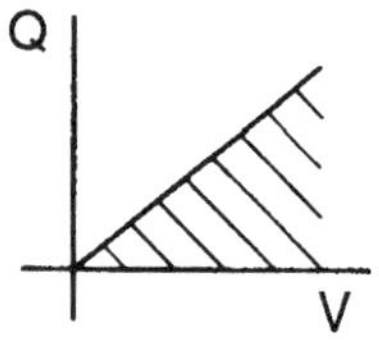

Memorise

6. $E = \frac{1}{2}QV$ $E = \frac{1}{2}\frac{Q^2}{C}$ $E = \frac{1}{2}CV^2$

Solve

7. Find the energy stored in each capacitor.

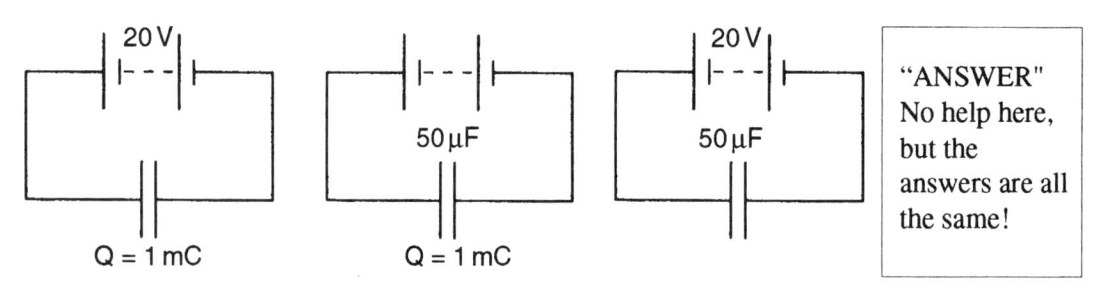

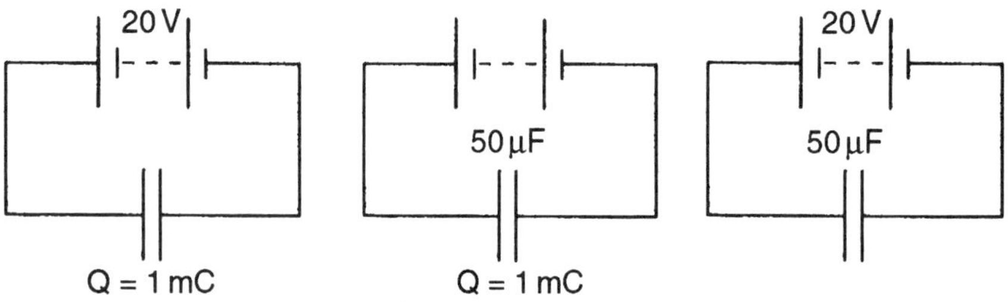

"ANSWER"
No help here, but the answers are all the same!

Memorise

8. Current/time and voltage/time graphs for an RC charging circuit.

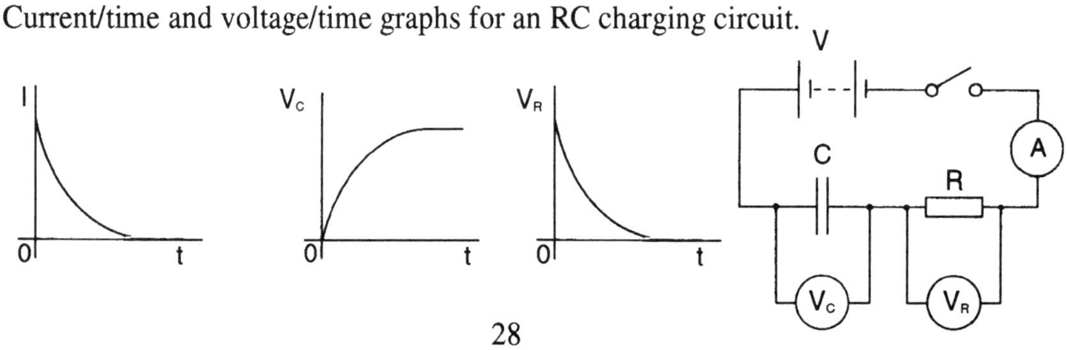

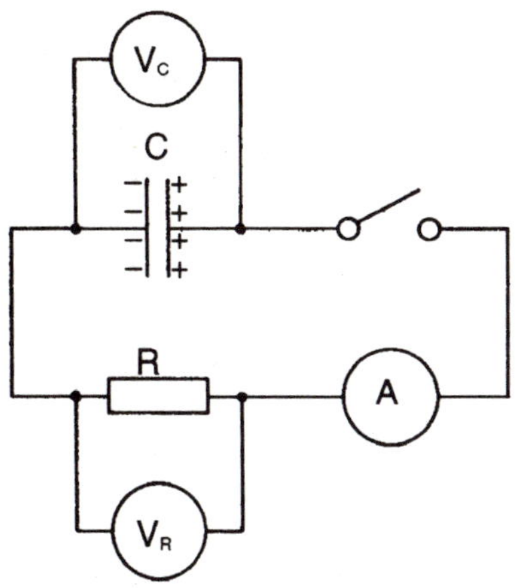

DISCHARGING CIRCUIT

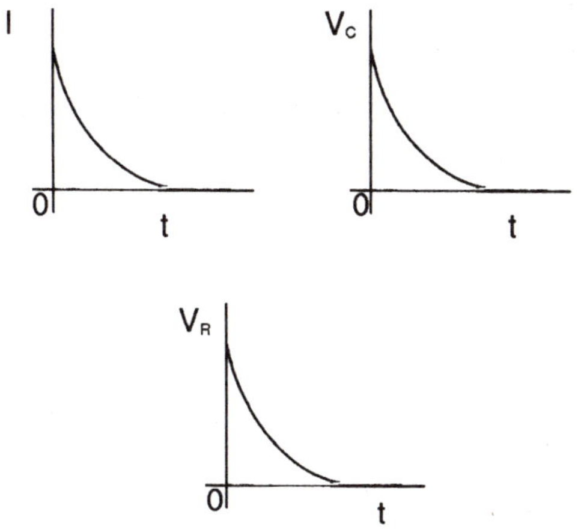

Understand

9. **Increasing** the **capacitance** will cause the **time** for charging (or discharging) to **increase**. **More charge** will be stored ($Q = CV$).

10. **Increasing** the **resistance** will cause the **time** for charging (or discharging) to **increase**.

Less current will flow at the start ($I = \frac{V}{R}$).

Solve

11.

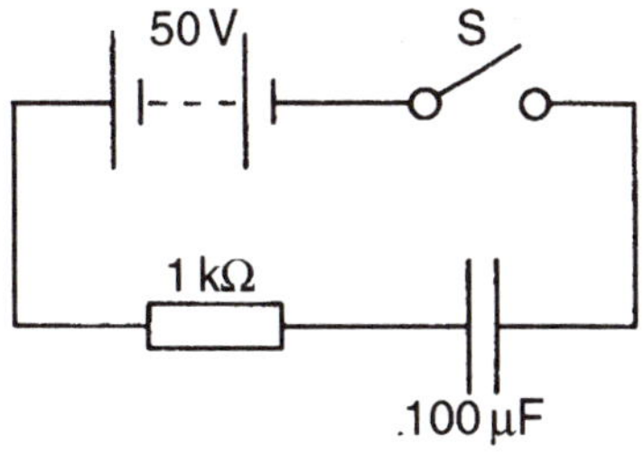

(a) Find:- (i) the initial current after S is closed.
(ii) the voltage across the capacitor when it is fully charged.
(iii) the voltage across the resistor at that time.
(iv) the final charge stored on the capacitor.

(b) How would the time taken to charge the capacitor be affected by:-
(i) increasing R (ii) reducing C (iii) increasing V?

ANSWERS

(a) (i) $I = \frac{V}{R} = \frac{50}{1000} = \mathbf{0.05\,A}$
(ii) **50 V**
(iii) **0 V**
(iv) $Q = CV = 100 \times 10^{-6} \times 50 = \mathbf{5 \times 10^{-3}\,C}$

(b) The time would (i) **increase** (ii) **reduce** (iii) **be unaffected**.

Understand

12. In a d.c. circuit, when a capacitor is fully charged, no more current flows. The **capacitor blocks d.c.**

13. In an a.c. circuit, the capacitor opposes the alternating current but does not block it completely. This opposition becomes less as the frequency increases. As a result, the current increases as the frequency increases.

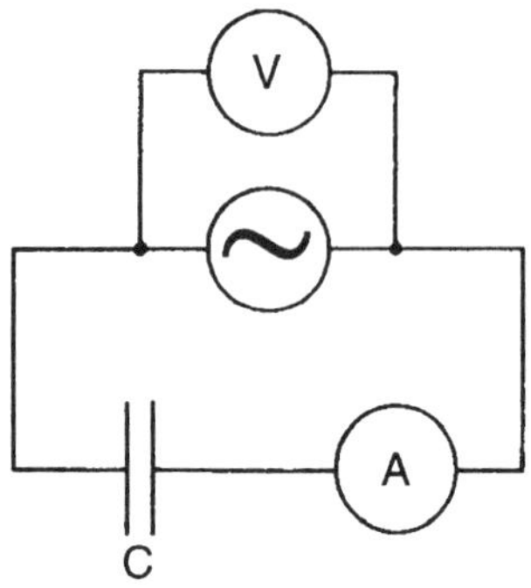

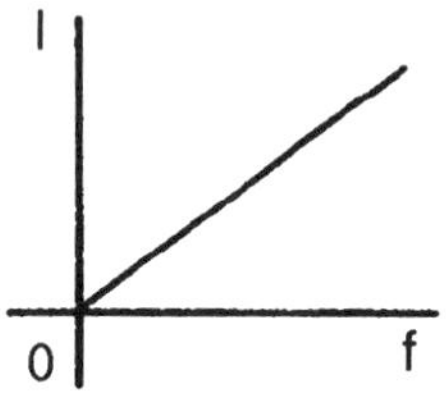

$I \propto f$

14. Capacitors have practical uses:-

(i) Storing energy $E = \frac{1}{2}QV$ to be released suddenly, e.g. in an electronic flashgun.

(ii) Blocking d.c. but allowing a.c. to pass.

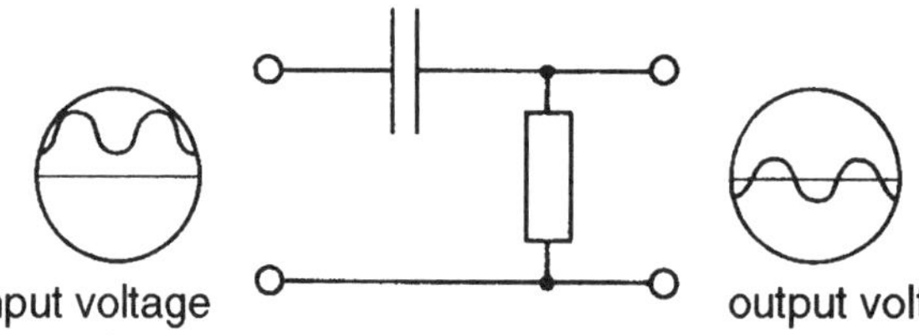

(iii) Smoothing a rectified waveform.

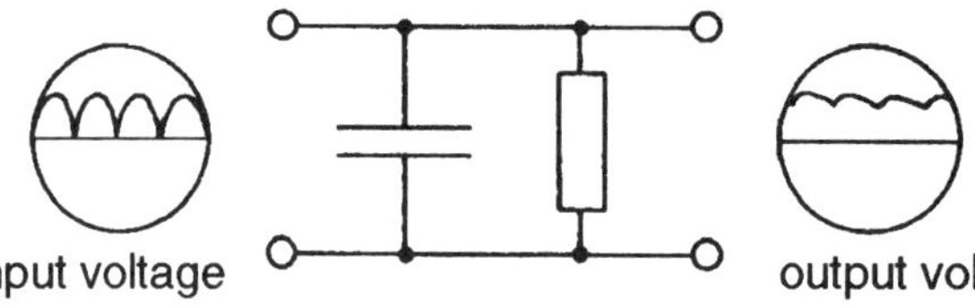

ANALOGUE ELECTRONICS

Understand

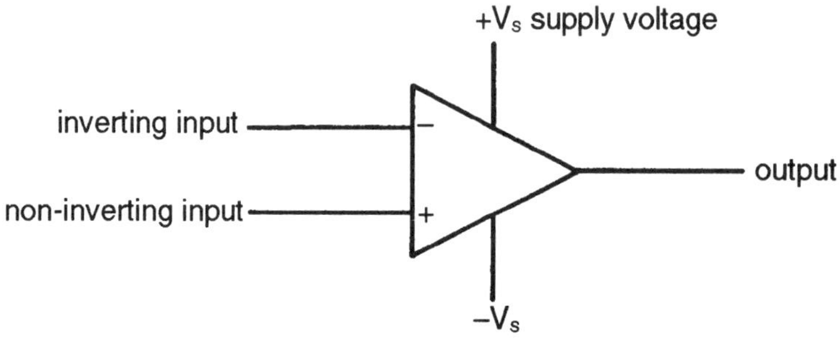

1. The operational amplifier (op. amp.) increases the voltage of an input signal.

State

In the **inverting mode**:-

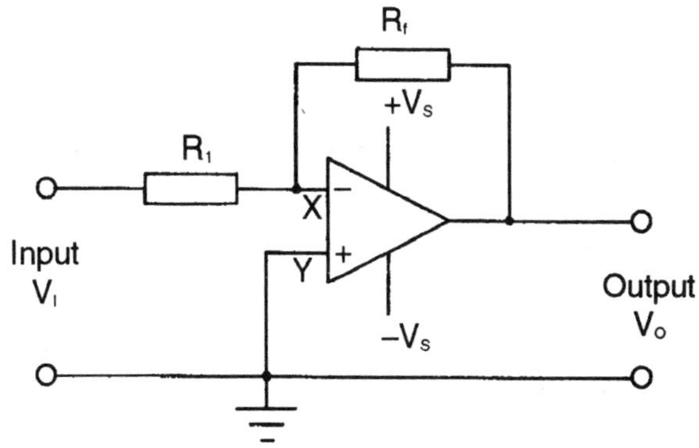

2. The op. amp. has an infinite input resistance. It draws no current from the source.

3. Both the inverting and the non-inverting inputs X and Y are at the same potential. In the diagram above, both these inputs are at zero (earth) potential.

4. The output voltage V_O can never exceed the supply voltage V_S.

5. If the input voltage, V_I, is positive, then the output voltage, V_O, is negative (and vice versa).

Memorise

6.

$$\text{Gain} = \frac{R_f}{R_I} = \frac{-V_O}{V_I}$$

Solve

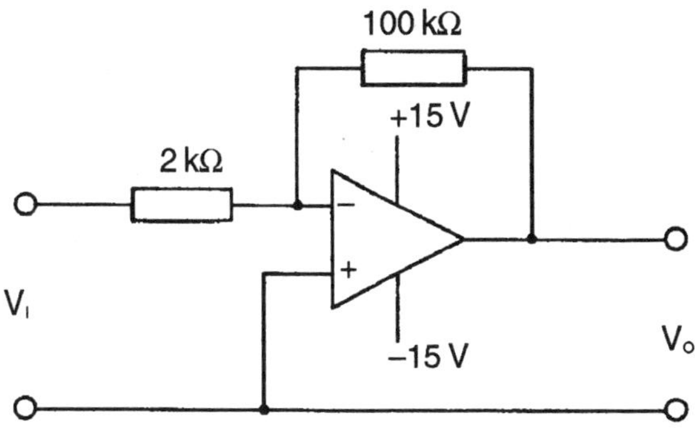

7. (a) Find the gain of this op. amp.

(b) Sketch a graph showing how the output voltage varies with time for each of the following input voltages.

(i)

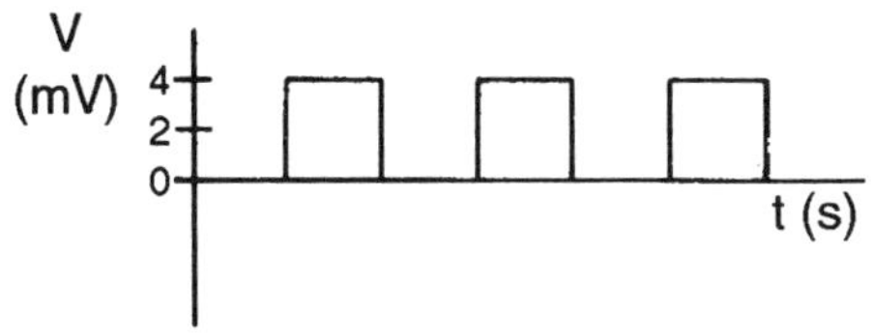

(ii)

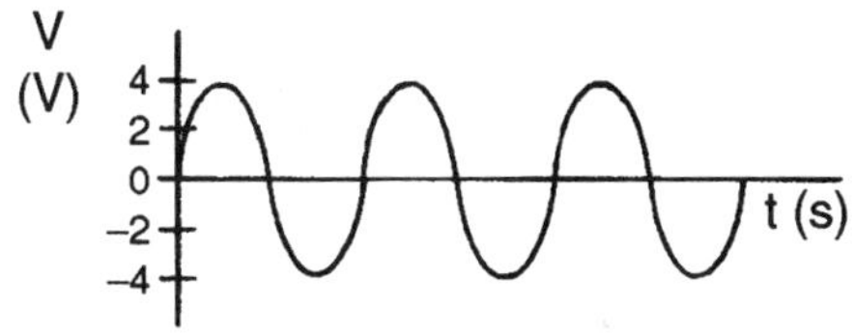

ANSWERS

(a) $\text{Gain} = \frac{R_f}{R_I} = \frac{100 \times 10^3}{2 \times 10^3} = \mathbf{50}$

(b) (i)

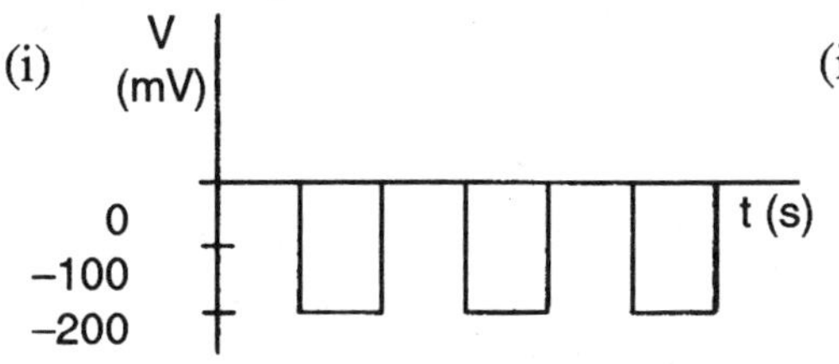

(ii)

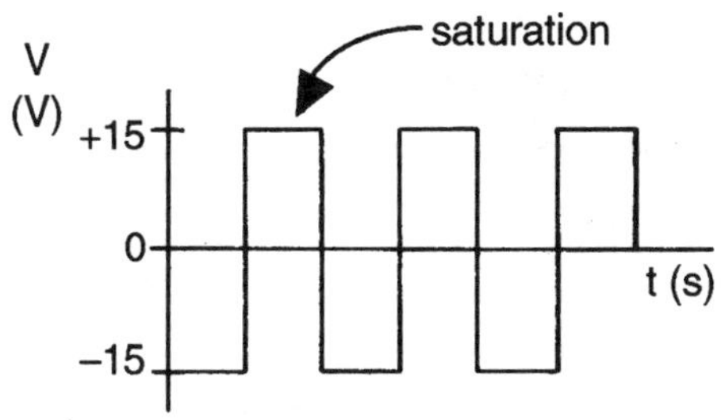

State

8. In the differential mode, the op. amp. amplifies the potential difference between the two inputs, i.e. $(V_2 - V_1)$.

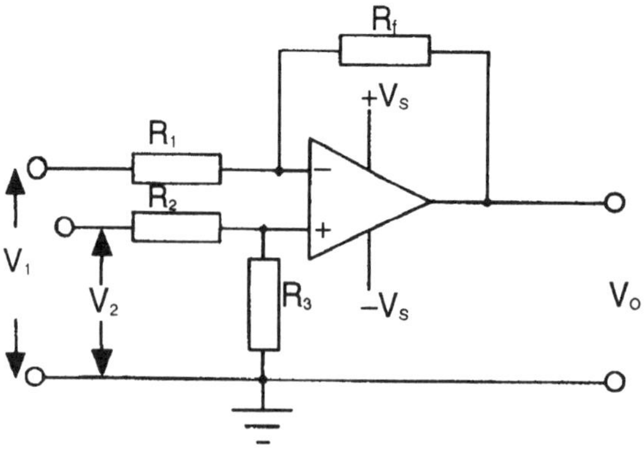

Memorise

9. If $\frac{R_f}{R_1} = \frac{R_3}{R_2}$, then

$$V_O = (V_2 - V_1)\frac{R_f}{R_1}$$

Solve

10. Find V_O in this circuit.

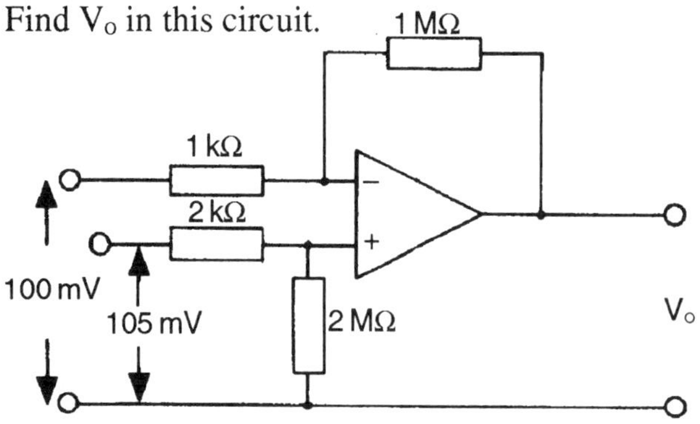

ANSWER

Check first, is $\frac{R_f}{R_1} = \frac{R_3}{R_2}$?

Yes, so apply the formula,

$$V_O = (105 - 100) \times 10^{-3} \times \frac{10^6}{10^3}$$

$$= 5 \times 10^{-3} \times 10^3 = \mathbf{5\,V}$$

Understand

11. Op. amps. can be connected to Wheatstone Bridges to amplify small out of balance voltages.

12. Op. amps. often can't provide enough current to light a bulb or sound a buzzer. They can be connected to transistors which will do the job.

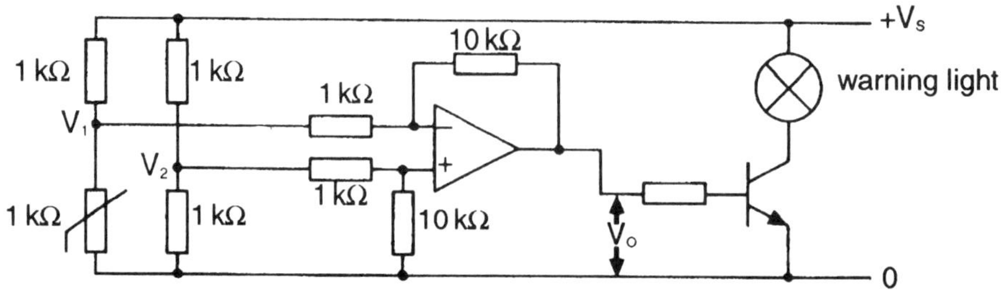

This circuit responds to very small increases in temperature by lighting up the warning light ⊗. The thermistor resistance is 1 kΩ so the Wheatstone Bridge is balanced, $V_1 = V_2 \Rightarrow V_O = 0$. If there is a small increase in temperature, the thermistor resistance decreases. This makes V_1 slightly less than V_2. The op. amp. amplifies $(V_2 - V_1)$ to produce a larger output voltage V_O which "switches on" the transistor. The transistor allows a big enough current through the bulb to light it up.

4. RADIATION AND MATTER

WAVE TERMS

Memorise

1. Period $= \dfrac{1}{\text{frequency}}$ $\boxed{T = \dfrac{1}{f}}$ Frequency $= \dfrac{1}{\text{period}}$ $\boxed{f = \dfrac{1}{T}}$

State

2. The frequency of a wave is the same as the frequency of its source (provided the source doesn't move).

3. The energy of a wave depends on its amplitude, **a**.

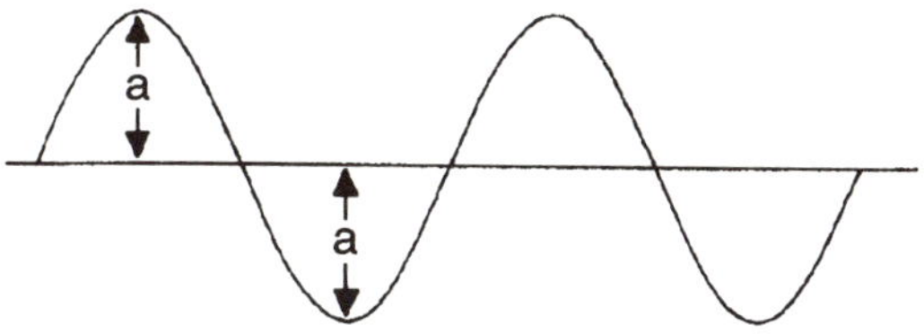

Understand

4. The meaning of **phase**:

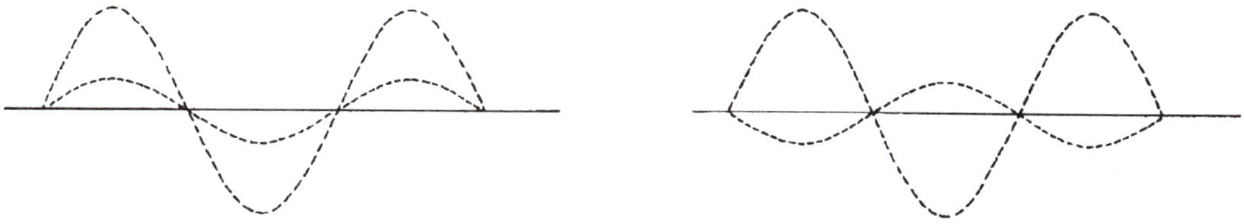

two waves in phase

two waves completely out of phase

5. If waves coming from two sources have the same frequency, wavelength and speed and are in phase, then the two sources are **coherent**.

Solve

6. A source produces a complete wave every $\dfrac{1}{200}$ s. Find T and f.

ANSWER $T = \dfrac{1}{200}\,\text{s} = \mathbf{0.005\,s}$ $f = \dfrac{1}{T} = \dfrac{1}{0.005} = \mathbf{200\,Hz}$

7. Which wave: (a) has the longest period
(b) is in phase with wave B
(c) has the lowest frequency
(d) has the smallest amplitude?

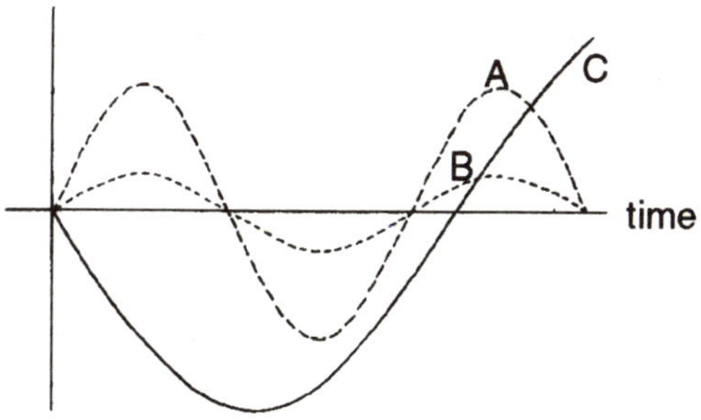

ANSWERS (a) **C** (b) **A** (c) **C** (d) **B**

INTERFERENCE AND DIFFRACTION

Understand

1. **Constructive Interference** occurs when waves from two coherent sources meet **in phase** to produce a wave of double the amplitude. They can meet in phase as follows:-

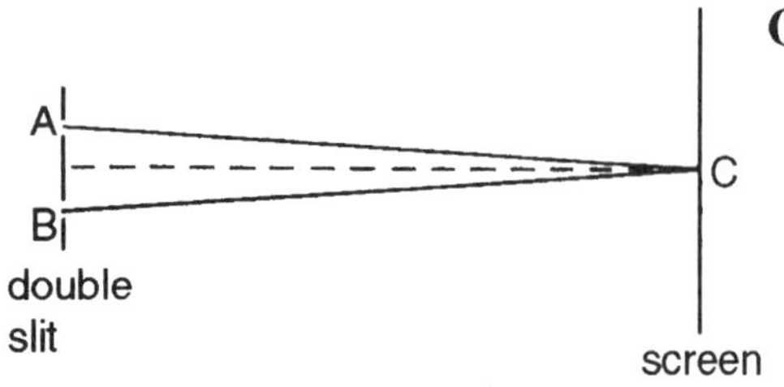

The two wave trains travel equal distances (AC = BC), i.e. there is no path difference.

OR

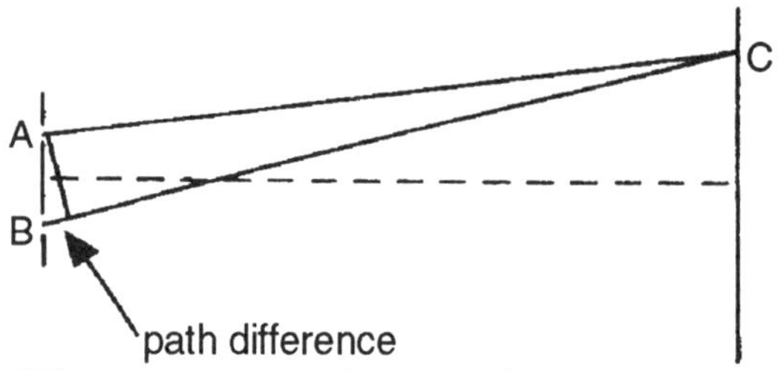

The two wave trains travel different distances. The path difference = λ, 2λ, 3λ, etc.

Solve

2. A microwave detector picks up a second maximum at point C. If distance AC = 100 cm, find distance BC.

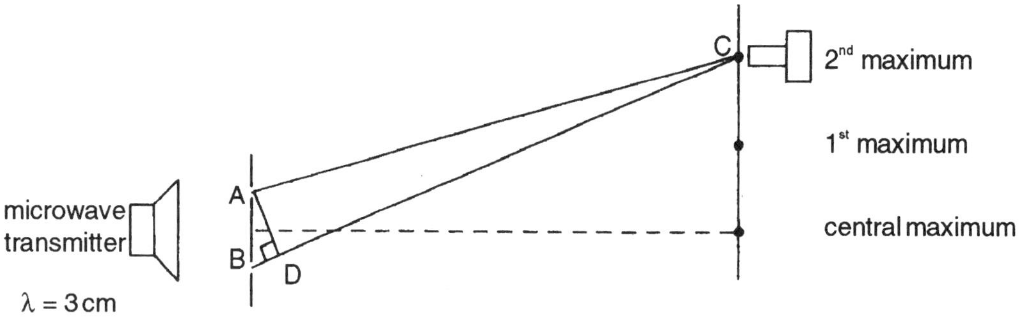

ANSWER

For a second maximum, the path difference (BD) $= 2\lambda$

$= 6\text{ cm}$

$\Rightarrow$ **BC = 106 cm.**

Understand

3. **Destructive Interference** occurs when waves from two coherent sources meet completely out of phase and cancel each other out.

For destructive interference, the path difference = $\frac{1}{2}\lambda$, $1\frac{1}{2}\lambda$, $2\frac{1}{2}\lambda$, etc.

Memorise

4. The diffraction grating equation: For a maximum, $n\lambda = d \sin\theta$

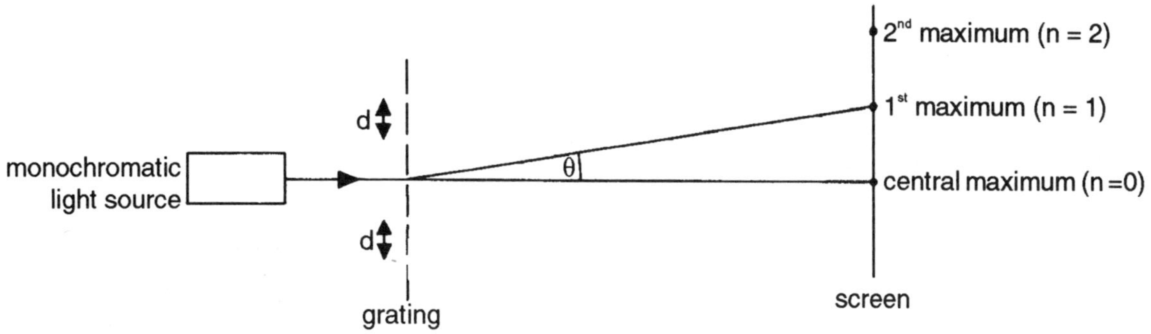

Understand

5. d is the distance between the slits on the grating. It is usually very small, e.g. 10^{-6}m.

6. You can find d if you are given N, the number of slits (or lines) per metre on the grating. $d = \frac{1}{N}$

<u>Example</u>

$$\begin{aligned} N &= 500 \text{ lines per mm} \\ \Rightarrow N &= 500 \times 1000 \text{ lines per m} \\ &= 5 \times 10^5 \text{ lines per m} \\ \Rightarrow d &= \frac{1}{N} \\ &= \frac{1}{5 \times 10^5} \\ &= \mathbf{2 \times 10^{-6}\,m} \end{aligned}$$

Solve

7. A laser produces a diffraction pattern using a grating of spacing 2×10^5 lines per metre. The first maximum is detected at angle θ of 20° as shown.

(a) Find the wavelength of the laser radiation.
(b) Would you be able to see it?

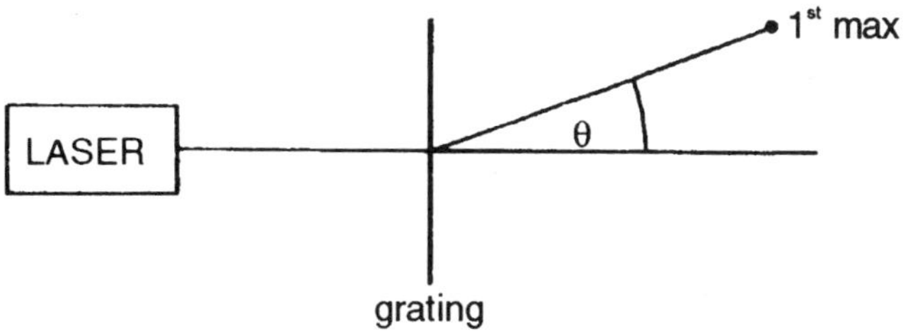

ANSWERS

(a) $n\lambda = d \sin\theta$ and $n = 1$ (1st max),

$$d = \frac{1}{2 \times 10^5} = 5 \times 10^{-6}\text{ m}, \quad \theta = 20°$$

$$\Rightarrow \lambda = 5 \times 10^{-6} \times \sin 20° = \mathbf{1.71 \times 10^{-6}\ m} \ (= 1710\text{ nm})$$

(b) This wavelength is in the infrared. You could not see it.

8. Wavelengths of light (λ) are often quoted in nanometres, nm.

1 nm = 10^{-9} m.

Memorise

9. Approximate values for the wavelengths of three colours:-

red	650 nm
green	500 nm
blue	450 nm

Describe

10. An experiment to measure the wavelength of a monochromatic light source using a grating. It's really just a combination of the diagram at the top of page 35 and the problem at the bottom.

11. The differences in white light spectra produced by
(a) a diffraction grating
(b) a prism

Grating Spectrum

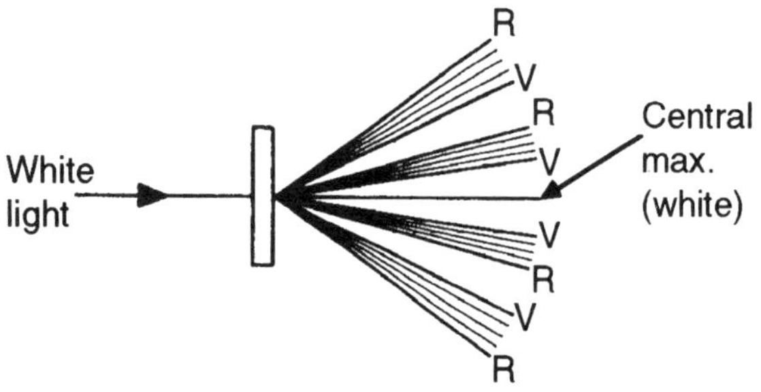

(i) Central maximum is white
(ii) Several coloured spectra
(iii) Red diffracted more than violet

Prism Spectrum

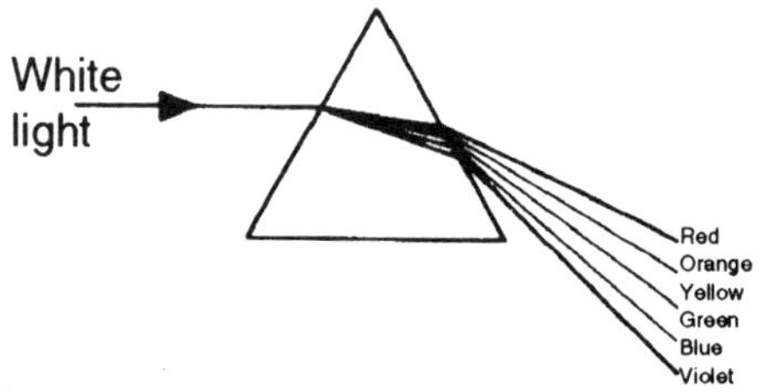

(i) One spectrum only
(ii) Red refracted less than violet

REFRACTION AND TOTAL INTERNAL REFLECTION

Memorise

1.

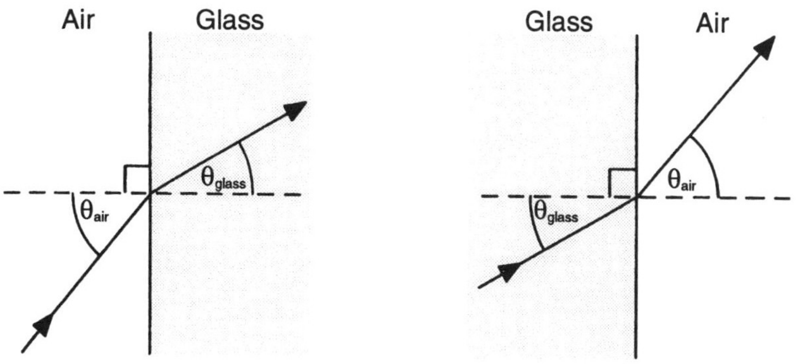

$$\frac{\sin \theta_{air}}{\sin \theta_{glass}} = n$$

n is called the absolute refractive index of glass.

State

2. The refractive index depends on the frequency of the light.

Solve

3. Find the absolute refractive index of the water.

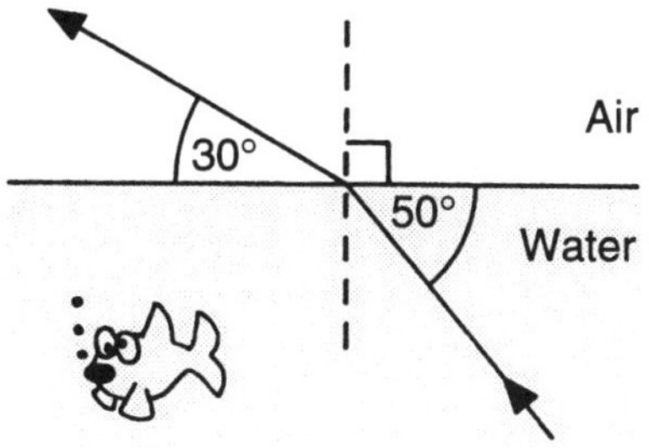

ANSWER

$\theta_{water} = 40°$

$\theta_{air} = 60°$

$$n = \frac{\sin \theta_{air}}{\sin \theta_{water}}$$

$$\Rightarrow \quad n = \frac{\sin 60°}{\sin 40°}$$

$$\Rightarrow \quad n = \frac{0.866}{0.643}$$

$$\Rightarrow \quad \mathbf{n = 1.35}$$

State

4. Wavelength λ and velocity v change during refraction, but **frequency does not change**.

Memorise

5. When light passes from medium 1 to medium 2

$$\frac{\sin \theta_1}{\sin \theta_2} = \frac{\lambda_1}{\lambda_2} = \frac{v_1}{v_2}$$

Solve

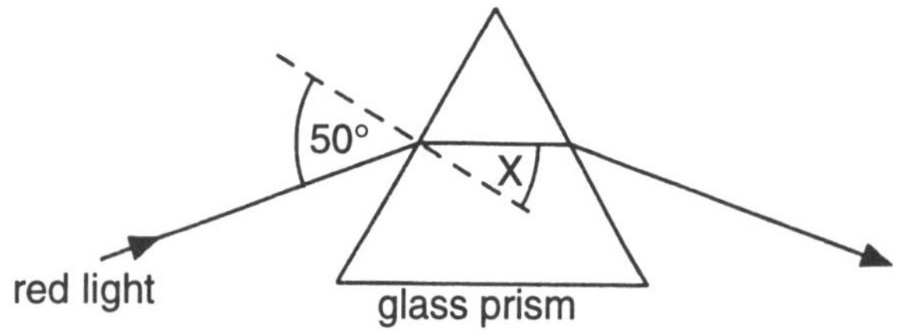

6. Red light of wavelength 650 nm passes from air into a glass prism as shown. The refractive index of the glass (for red light) is 1.52.

(a) Find:- (i) the angle X

(ii) the wavelength of the light in the glass.

(b) When blue light hits the prism at the same angle of incidence (50°), angle X is found to be 29.6°. Find the refractive index of the glass for blue light.

ANSWERS

(a) (i) $\dfrac{\sin\theta_{air}}{\sin\theta_{glass}} = 1.52$

$\Rightarrow \sin\theta_{glass} = \dfrac{\sin\theta_{air}}{1.52}$

$\Rightarrow \sin X = \dfrac{\sin 50°}{1.52}$

$\Rightarrow$ **X = 30.3°**

(a) (ii) $\dfrac{\sin\theta_{air}}{\sin\theta_{glass}} = \dfrac{\lambda_{air}}{\lambda_{glass}}$

$\Rightarrow 1.52 = \dfrac{650}{\lambda_{glass}}$

$\Rightarrow$ λ_{glass} **= 428 nm**

(b) $\dfrac{\sin\theta_{air}}{\sin\theta_{glass}} = n \Rightarrow \dfrac{\sin 50°}{\sin 29.6°} = n \Rightarrow$ **n = 1.55**

Understand

7. Total internal reflection occurs when the angle of incidence in the 'denser' medium (e.g. glass) is greater than the critical angle, θ_C.

8. The critical angle is reached when the angle of refraction is 90°.

9. These three diagrams show what happens as the angle θ in the glass is increased:

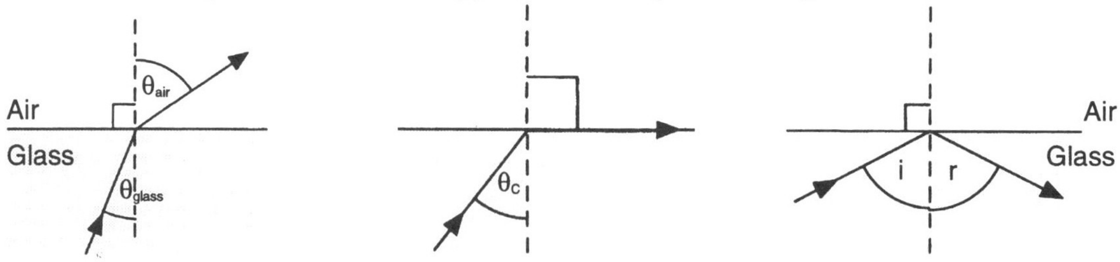

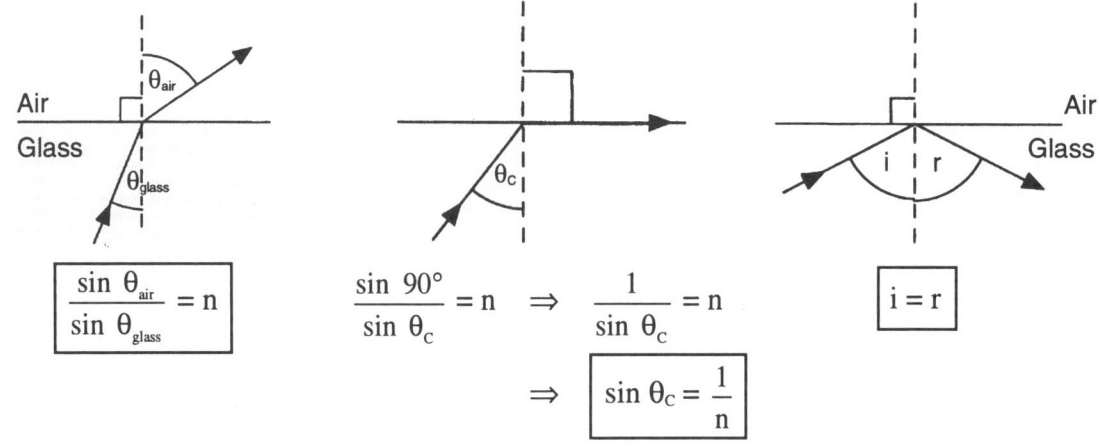

$\dfrac{\sin\theta_{air}}{\sin\theta_{glass}} = n$

$\dfrac{\sin 90°}{\sin\theta_C} = n \Rightarrow \dfrac{1}{\sin\theta_C} = n$

$\Rightarrow \sin\theta_C = \dfrac{1}{n}$

$i = r$

Memorise

10. $\sin\theta_C = \dfrac{1}{n}$

Solve

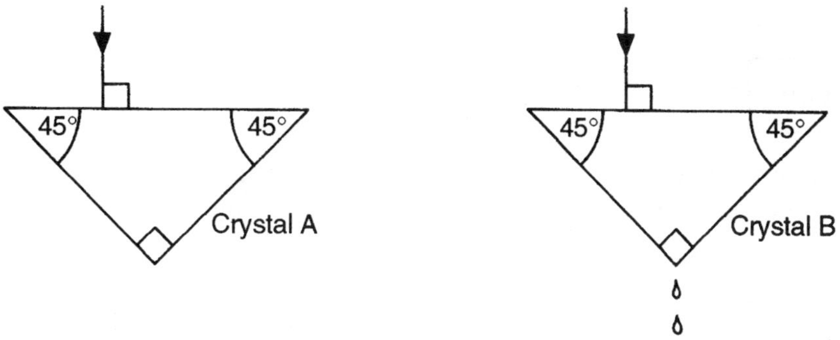

11. Crystal A is a diamond (n = 2.4). Crystal B is ice (n = 1.3). Copy the two diagrams and show clearly the complete paths of the light rays through the two crystals.

ANSWER

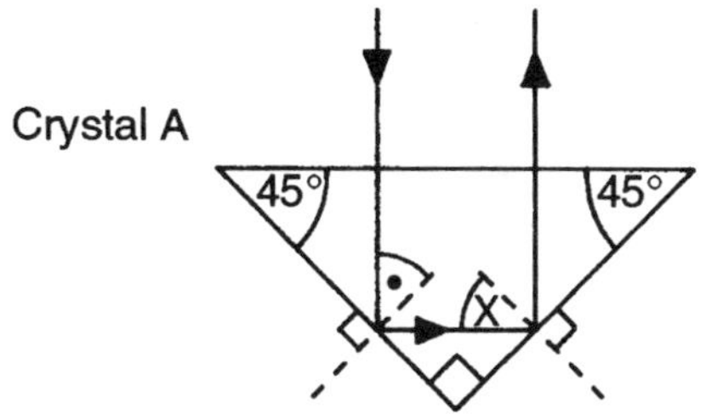

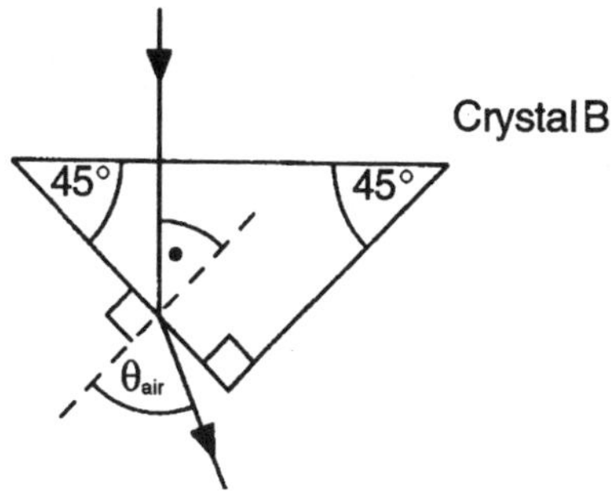

$$\sin\theta_C = \frac{1}{n} = \frac{1}{2.4} \Rightarrow \theta_C = \mathbf{24.6°}$$

The • angle is 45°. This is greater than 24.6° so total internal reflection occurs (i = r). Angle X is also 45° so again we have total internal reflection. The ray finally emerges as shown.

$$\sin\theta_C = \frac{1}{n} = \frac{1}{1.3} \Rightarrow \theta_C = \mathbf{50.3°}$$

The • angle is 45°. This is less than 50.3° so refraction occurs.

$$\frac{\sin\theta_{air}}{\sin\theta_{ice}} = 1.3 \Rightarrow \frac{\sin\theta_{air}}{\sin 45°} = 1.3$$

$$\Rightarrow \theta_{air} = \mathbf{66.8°}$$

LIGHT INTENSITY

Memorise

1. $\text{Intensity} = \dfrac{\text{Power}}{\text{Area}}$ $\boxed{I = \dfrac{P}{A}}$ Units of intensity: $W\,m^{-2}$

2. Intensity varies inversely as $(\text{distance})^2$ from a point source.

$$\boxed{I \propto \frac{1}{d^2}} \quad \boxed{I = \frac{k}{d^2}}$$ This is an **inverse square law**.

Describe

3. An experiment to prove the inverse square law for light from a point source. Check it out now!

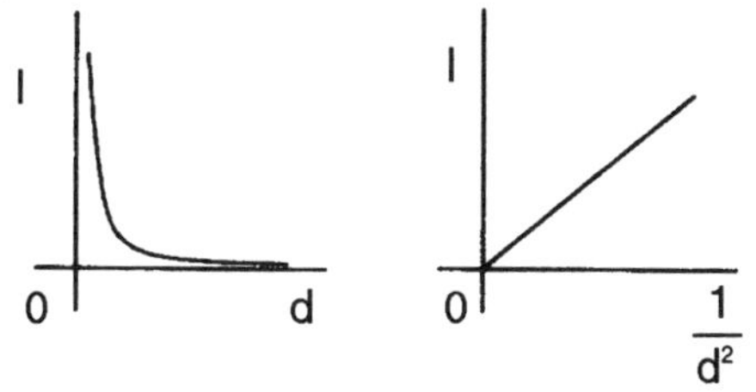

THE PHOTOELECTRIC EFFECT

Understand

1. In some experiments (e.g. interference), light and the other forms of electromagnetic radiation behave like continuous waves of energy, but in the photoelectric effect they behave like individual "lumps" or "packets" of energy called photons.

2. In the photoelectric effect, electrons are knocked off a metal surface by photons. The electrons can form a photoelectric current.

Memorise

3. The energy of a single photon $\boxed{E = hf}$

4. $\boxed{\text{Intensity of a beam of photons} = Nhf}$ where N = number of photons per second hitting a square metre.

Understand

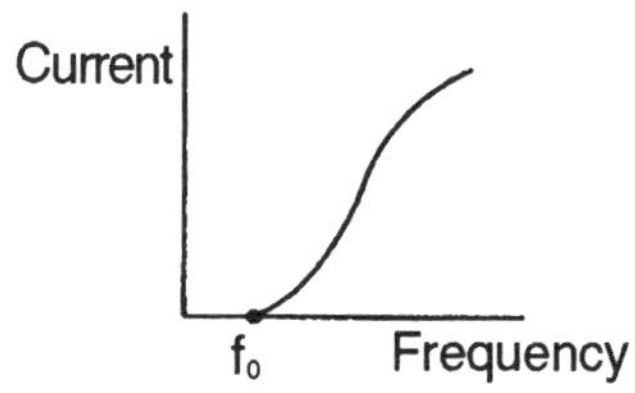

5. Below the threshold frequency f_0, there is no photoelectric current. Increasing the intensity of the radiation below f_0 makes absolutely no difference.

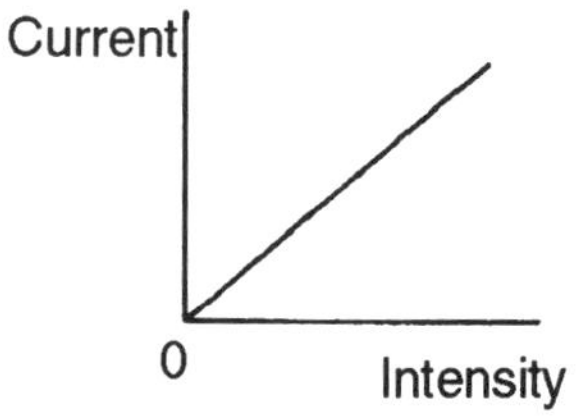

6. Above the threshold frequency, there is a photoelectric current. The current is proportional to the intensity. In simple language, the more photons that hit the metal every second the more electrons are knocked off the surface every second. (Think of each photon knocking an electron off the metal.)

7. The minimum energy needed to knock an electron off a particular metal surface is called the Work Function, W, of the metal and is equal to hf_0. $\boxed{W = hf_0}$

8. If photons have energy greater than hf_0, the extra energy appears as the kinetic energy of the electrons. $\boxed{hf = W + \frac{1}{2}mv^2}$

Solve

9. A metal has a work function of 5×10^{-19} J. Can the following electromagnetic radiations eject an electron from the metal?
   (a) Ultra violet photons ($f = 10^{15}$ Hz); (b) Radio 3 photons ($f = 92$ MHz)?
   (Planck's constant, $h = 6.63 \times 10^{-34}$ J s)

ANSWERS

(a) For uv: $E = hf = 6.63 \times 10^{-34} \times 10^{15} = 6.63 \times 10^{-19}$ J **Yes!** (and the electrons would have kinetic energies of 1.63×10^{-19} J).

(b) For Radio 3: $E = hf = 6.63 \times 10^{-34} \times 9.2 \times 10^{7} = 6.1 \times 10^{-26}$ J **No**

SPECTRA

Understand

1. Electron energies in an atom are **quantised**. Electrons can only occupy certain definite energy levels. When an electron is bound to the nucleus, its energy has a negative value. (An electron just freed from the atom by ionisation would have a zero energy value. If it moved away from the atom, it would have a positive kinetic energy.)

2. Make sure you know the meanings of the terms:- **ground state, excited state, ionisation level, electron transition, emission line, absorption line.**

Look at this worked example to help you:-

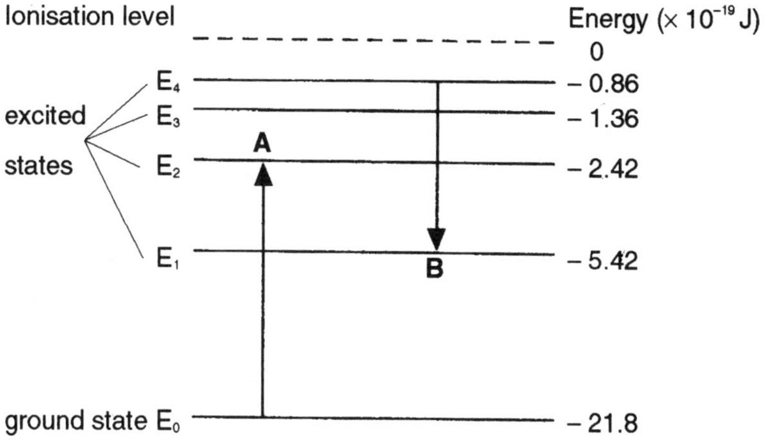

The diagram shows two possible electron transitions **A** and **B** for the electron in a hydrogen atom. Which one would produce an absorption line and which one an emission line? Find the frequency and wavelength of each line. ($h = 6.63 \times 10^{-34}$ J s, speed of light $v = 3 \times 10^{8}$ m s^{-1})

ANSWERS

A produces an absorption line.

$$E = hf$$

$$\Rightarrow (21.8 - 2.42) \times 10^{-19} = 6.63 \times 10^{-34} \times f$$

$$\Rightarrow \quad \mathbf{f = 2.92 \times 10^{15}\ Hz}$$

$$\lambda = \frac{v}{f} = \frac{3 \times 10^{8}}{2.92 \times 10^{15}} = \mathbf{1.03 \times 10^{-7}\ m}$$

B produces an emission line.

$$E = hf$$

$$\Rightarrow (5.42 - 0.86) \times 10^{-19} = 6.63 \times 10^{-34} \times f$$

$$\Rightarrow \quad \mathbf{f = 6.88 \times 10^{14}\ Hz}$$

$$\lambda = \frac{v}{f} = \frac{3 \times 10^{8}}{6.88 \times 10^{14}} = \mathbf{4.36 \times 10^{-7}\ m}$$

Describe

3. Find your notes or a textbook and check up how you would produce:-
   (a) a continuous emission spectrum
   (b) a line emission spectrum
   (c) an absorption spectrum.
   Finally, find out why the Sun's spectrum contains absorption lines.

Understand

1. **Spontaneous emission**

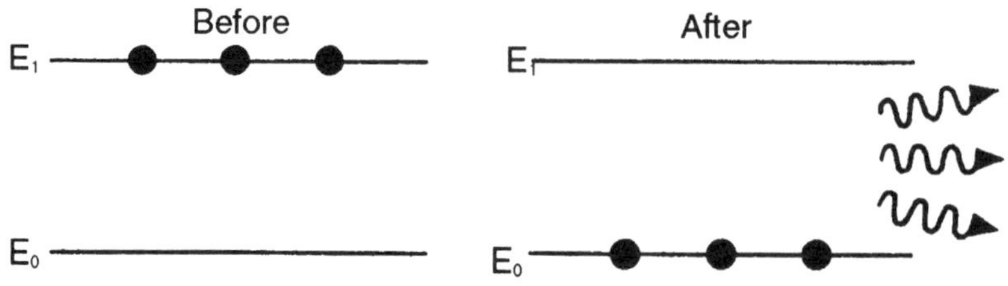

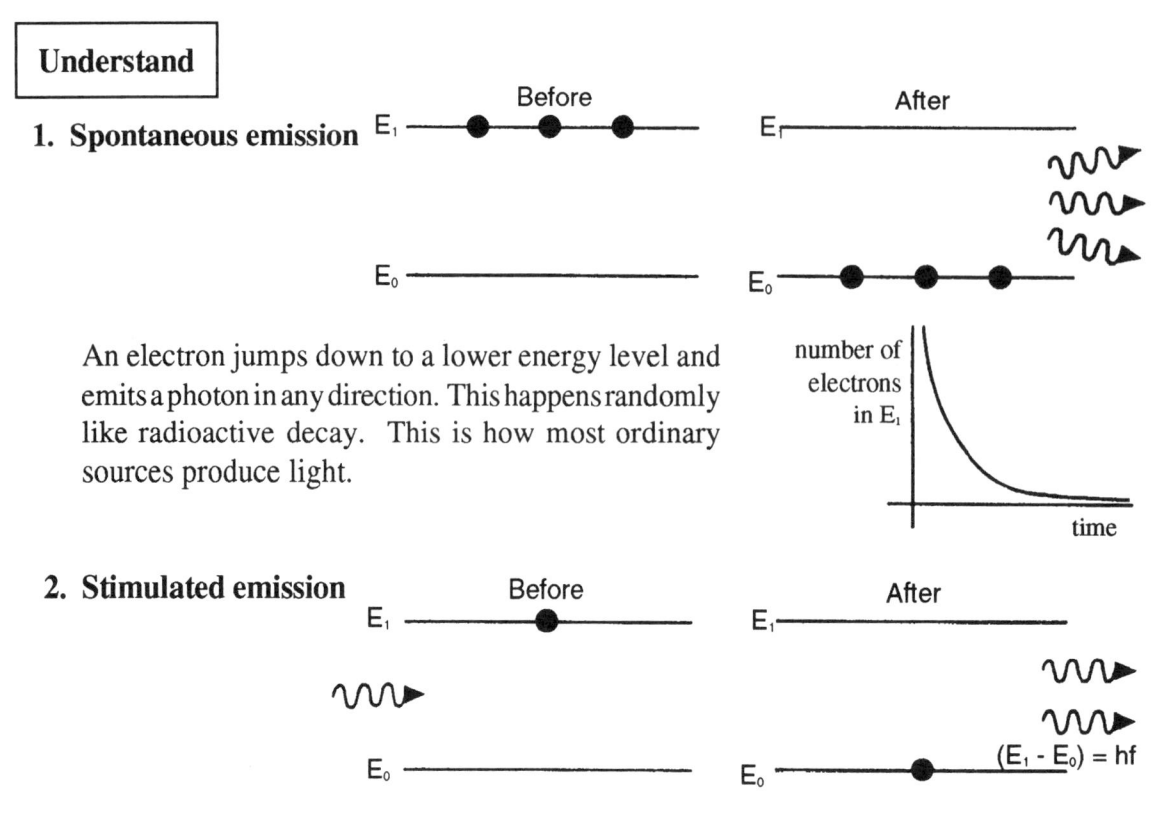

An electron jumps down to a lower energy level and emits a photon in any direction. This happens randomly like radioactive decay. This is how most ordinary sources produce light.

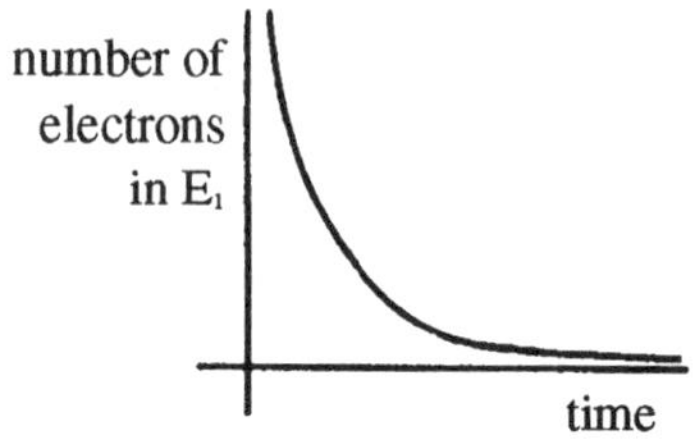

2. **Stimulated emission**

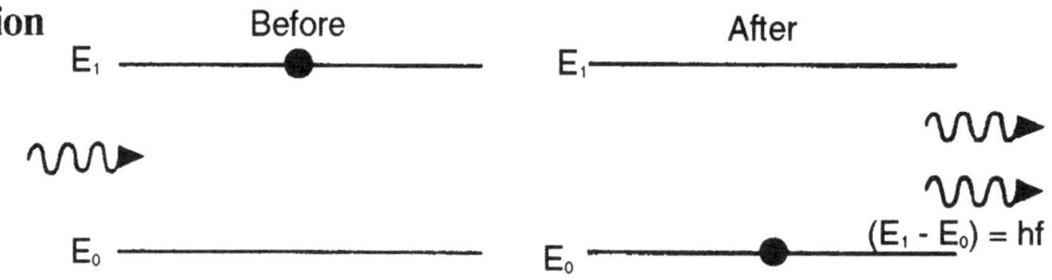

An incoming photon "encourages" an electron to jump down to a lower energy level and emit a photon. The two photons are in phase and moving parallel to each other.

3. **Stimulated absorption**

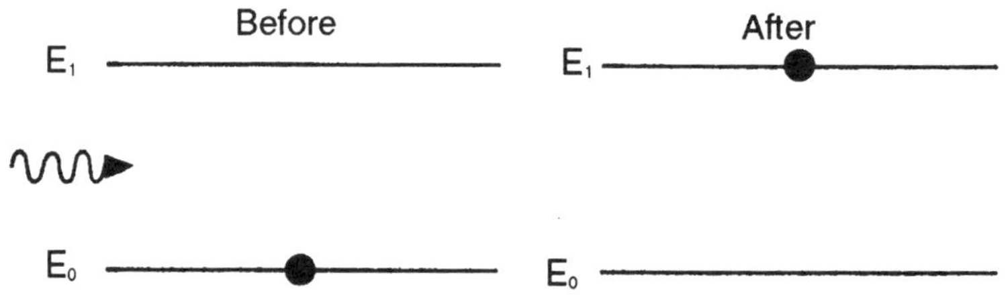

Incoming photons are "absorbed" by electrons in a lower energy level. The electrons jump to a higher level.

4. In normal crystals and gases **most electrons** are in the **lower energy levels**, so if a photon happens to come along the chances are that it will cause stimulated absorption.

5. The "secret" of the laser is to produce a situation where **most electrons** are in the **higher energy levels (a population inversion)**. So now the chances are that an incoming photon will cause stimulated emission.

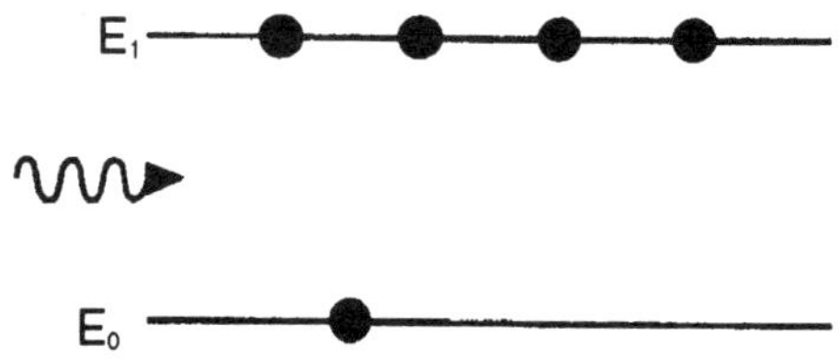

6. The purpose of the Laser's two parallel **mirrors**:-

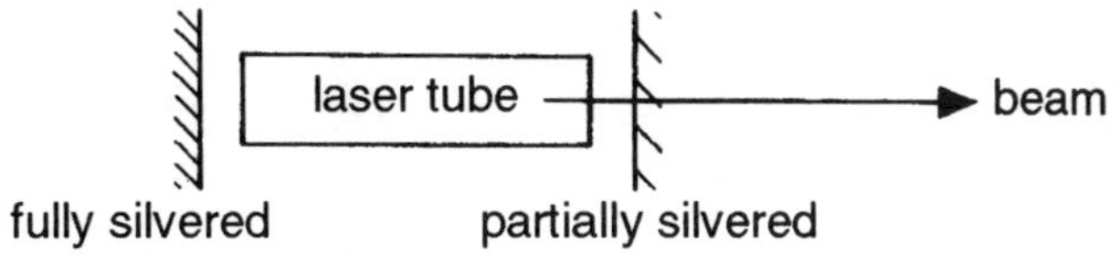

The photons are reflected back and forward along the tube many times, increasing the amount of stimulated emission. Only photons which are travelling straight along the tube continue to be reflected. This makes the laser beam intense and parallel.

7. The laser beam may cause eye damage because of its **high intensity**.

<u>Example</u> A 2 mW laser which produces a 2 mm radius 'spot' on a screen

has an intensity $= \frac{\text{Power}}{\text{Area}} = \frac{2 \times 10^{-3}}{\pi \times (2 \times 10^{-3})^2} = \mathbf{159\,W\,m^{-2}}$.

That could be as much as twenty times greater than the intensity of the light on this page.

SEMICONDUCTOR DEVICES

State

1. Materials belong roughly to three categories depending on their electrical resistance:- **conductors**, **insulators** and **semiconductors**.

Memorise

2. One example of each:-
Conductor - Copper Insulator - Polythene Semiconductor - Silicon

Understand

3. **Doping** means adding a very few "impurity" atoms to a pure semiconductor. This **reduces** its **resistance**.

4. Doping can produce **N type semiconductor**. The majority of charge carriers are negative (electrons).

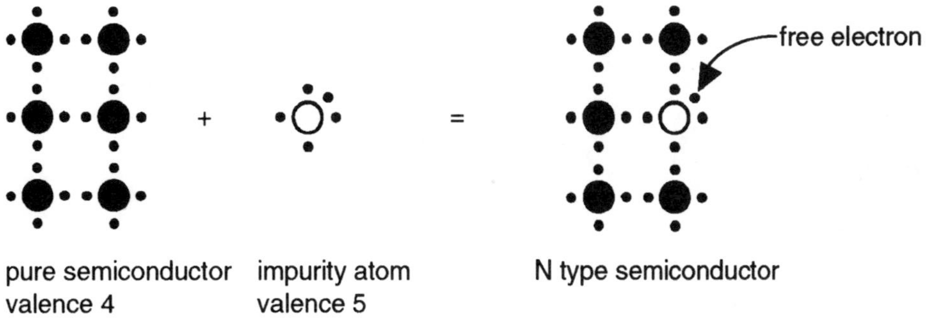

5. Doping can also produce **P type semiconductor**. The majority of charge carriers are positive (holes).

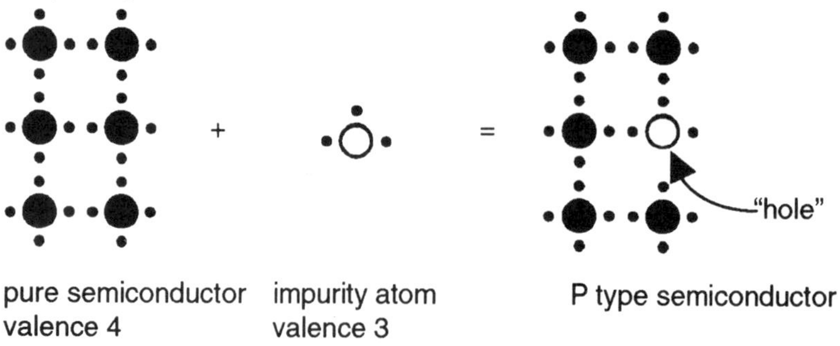

6. A P-N junction diode is made from a crystal of semiconductor material half of which is P-type and half N-type.

7. When the diode is **reverse biased**, almost **no current** flows.

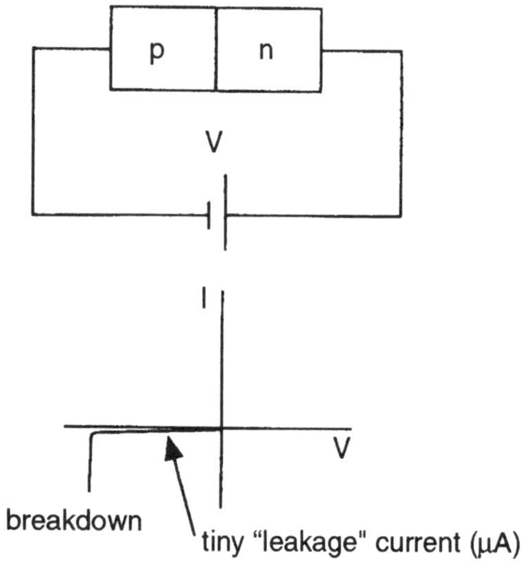

When the diode is **forward biased**, **current flows**.

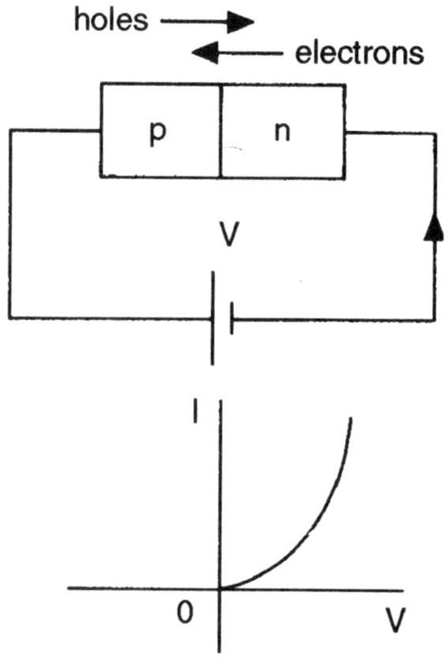

State

8. When a **light emitting diode** 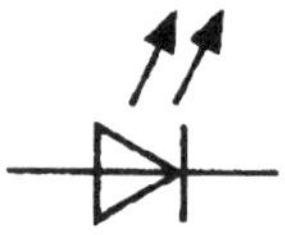 is forward biased, holes and electrons meet and recombine to give off energy in the form of photons.

9. In the **photodiode**, 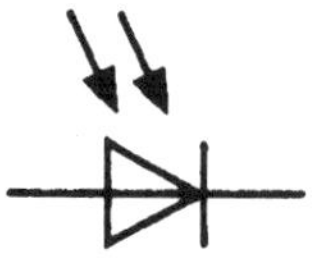 electrons and holes are produced when light hits the P-N junction.

Understand

10. The photodiode can be used in two different modes:-

Photovoltaic Mode

When light falls on the photodiode, pairs of electrons and holes are formed. They create an emf. The photodiode acts like a cell or battery. It's sometimes called a solar cell.

Photoconductive Mode

(a) The leakage current is independent of the cell voltage.
(b) The leakage current is directly proportional to the light intensity.
(c) Changes in light intensity produce almost instant changes in current.

The diode behaves like a light dependent resistor.

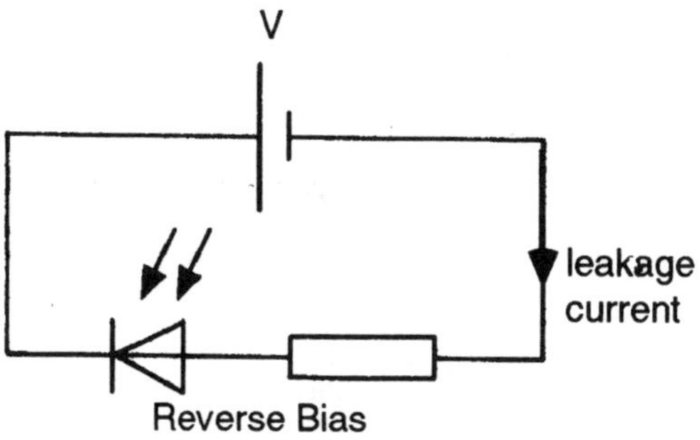

RADIOACTIVITY

Describe

1. The experiment that Rutherford did with alpha particles to find the structure of the atom. Look for the details in your notes or your textbook. Make some new notes, now!

Memorise

2.

Particle	*Nature*	*Charge*	*Absorption*	*Ionising Power*
α	Helium nucleus ^4_2He	+2	Paper	Very strong
β	Electron $^0_{-1}e$	–1	A few mm aluminium	Moderate
γ	Electromagnetic photon	0	A few cm lead	Weak

Understand

3. The symbol for a nuclide

Number of (protons + neutrons) —— $^{222}_{86}Rn$ —— Chemical symbol
Number of protons —— 86

This nuclide is radon. It has 86 protons and 136 neutrons (222 – 86 = 136).

Solve

4. Find the unknown quantities X and Y.

(a) $^{212}_{83}Bi \longrightarrow ^{208}_{81}Tl + X$ (b) $^{218}_{84}Po \longrightarrow ^{218}_{85}At + Y$

ANSWERS

(a) X mass = 4, charge = 2
It is an alpha particle ^4_2He.

(b) Y mass = 0, charge = –1
It is a beta particle $^0_{-1}e$.

5. The full equation in **4.** (a) is actually $^{212}_{83}Bi \longrightarrow ^{208}_{81}Tl + ^{4}_{2}He + Z$.
What is Z?

ANSWER

Z mass = 0, charge = 0 **It is a γ ray**.

Memorise

6. $\text{Activity} = \dfrac{\text{Number of nuclei decaying}}{\text{time}}$ $A = \dfrac{N}{t}$

7. Activity is measured in becquerels, Bq. 1 Bq = 1 decay per second.

8. $\text{Absorbed dose} = \dfrac{\text{Energy absorbed}}{\text{Mass of absorber}}$ $D = \dfrac{E}{m}$

9. Absorbed dose is measured in grays, Gy. 1 Gy = 1 joule per kilogram.

Solve

10. A source has an activity of 5 kBq. How many decays occur in a minute?

ANSWER

$A = \dfrac{N}{t} \Rightarrow N = At = 5000 \times 60 =$ **300 000 decays**

11. A 70 kg man absorbs 1.4 J of energy from a radioactive source. Find the absorbed dose.

ANSWER

$D = \dfrac{E}{m} = \dfrac{1.4}{70} =$ **0.02 Gy** [or 20 mGy]

Understand

12. Scientists have found that equal absorbed doses of different radiations don't do the same damage to us, e.g. 1 mGy of α particles does 20 times more damage to us than 1 mGy of β particles. To allow for this, each type of radiation is given a **quality factor Q**.

13. To allow for the differences between the radiations and to allow us to add the effects of different radiations together, a more useful unit is the Dose Equivalent, H.

Memorise

14. Dose equivalent = Absorbed dose × Quality factor

$$H = D \times Q$$

15. Dose equivalent is measured in sieverts, Sv.

16. Dose equivalent rate = $\dfrac{\text{Dose equivalent}}{\text{time}}$

$$\dot{H} = \frac{H}{t}$$

17. Dose equivalent rate is usually measured in millisieverts per year, $mSv\,yr^{-1}$ (or millisieverts per hour, $mSv\,h^{-1}$).

Solve

18. A worker receives 180 μGy of α radiation (Q = 20) and 120 μGy of β radiation (Q = 1) over 2 years. Calculate his dose equivalent rate.

ANSWER

$$\dot{H} = \frac{H}{t} = \frac{D \times Q}{t} = \frac{(180 \times 20) + (120 \times 1)}{2} = \frac{3720}{2} = \mathbf{1860\,\mu Sv\,yr^{-1}} \text{ (or } 1.860\,mSv\,yr^{-1})$$

State

19. The dose equivalent for members of the public should ideally not exceed 5 mSv in any one year. Radiation workers are allowed up to 50 mSv. This is in addition to normal background radiation ($2\,mSv\,yr^{-1}$).

Understand

20. Dose equivalent rate can be reduced by
(a) increasing distance from the source of radiation
(b) using a shielding material (an absorber).

Describe

21. An experiment to measure **half value thickness** of an absorber. This is an important experiment. Back to your notes or your textbook!

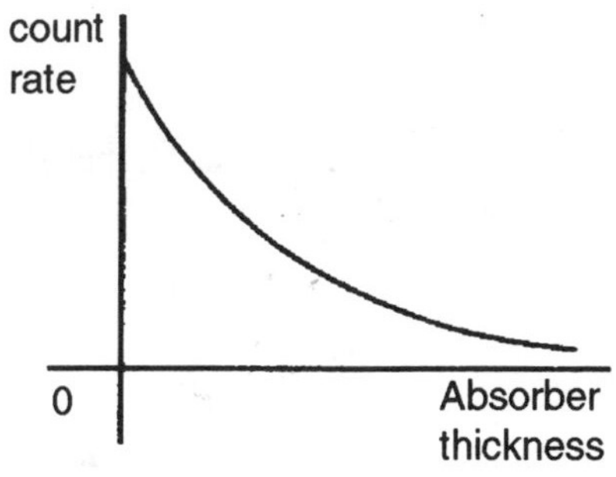

FISSION AND FUSION

State

1. **Fission** occurs when a large nucleus splits up into two smaller nuclei and several neutrons. Energy is released during fission.

2. **Induced fission** occurs when the large nucleus is bombarded by neutrons.

3. **Spontaneous fission** occurs without any outside influence.

Understand

4. The energy released in fission can be found from $E = mc^2$.

$1\ u = 1.660 \times 10^{-27}$ kg
$c = 3 \times 10^{-8}\ m\ s^{-1}$

Example $\quad {}^{235}_{92}U \quad + \quad {}^{1}_{0}n \quad \longrightarrow \quad {}^{141}_{56}Ba \quad + \quad {}^{92}_{36}Kr \quad + \quad 3\,{}^{1}_{0}n$

Mass before = 235.044 + 1.009 = 236.053

Mass after = 140.914 + 91.926 + 3.027 = 235.867

$\Rightarrow$ m = 236.053 – 235.867 = 0.186u

$E = mc^2$
$= 0.186 \times 1.660 \times 10^{-27} \times 9 \times 10^{16}$
$= \mathbf{2.779 \times 10^{-11}}$ **J** (This is the energy released by one single nucleus.)

State

5. **Fusion** occurs when two lighter nuclei join together to form a heavier nucleus. Energy is released during fusion.

Example $\quad {}^{3}_{1}H + {}^{2}_{1}H \longrightarrow {}^{4}_{2}He + {}^{1}_{0}n$

The energy released in fusion can be found from $E = mc^2$ as in **4.** above.

FOR MORE PRACTICE

For more practice at Higher Physics problem solving, Leckie & Leckie has also published *Higher Grade Physics Revision Questions* (ISBN 1-898890-30-7).